5,000 Years of Fine Cretan Wines

Glimpses of the History of Wine on a Beautiful Island

Frontispiece. Athenian red-figure krater (for mixing wine and water), showing a dinner party with music and wine for the host and his guests. Attributed to the Painter of the Louvre Centauromachy. Ca. 450 B.C., height 40.6 cm, diam. 38.1 cm. New York, The Metropolitan Museum of Art, gift of Sylvia de Cuevas, 2013, acc. 2013.158; https://www.metmuseum.org/art/collection/search/259196, accessed June 22, 2015; CC0 1.0, https://creativecommons.org/publicdomain/zero/1.0/.

5,000 Years of Fine Cretan Wines

―――――― • ――――――

Glimpses of the History of Wine on a Beautiful Island

Albert Leonard, Jr.
and
Philip P. Betancourt

Published by
INSTAP Academic Press
Philadelphia, Pennsylvania
2025

Editorial, Design, and Production: INSTAP Academic Press, Philadelphia, PA, USA
Project Editor: Susan Ferrence
Copyeditor: Tanya McCullough
Design and Production: Hilary Sperling
Cover Design: Molly Kaplan
Printing and Binding: Integrated Books International, Dulles, VA , USA, and CPI Group (UK) Ltd, Croydon, CR0 4YY

INSTAP Academic Press, a part of the Institute for Aegean Prehistory (INSTAP), was established to publish projects relevant to the history of the Aegean world, in particular from the Paleolithic to the 8th century B.C. It is a scholarly nonprofit publisher specializing in high-quality publications of primary source material from archaeological excavations as well as individual studies dealing with material from the prehistoric periods—exemplified by its *Prehistory Monographs* series of volumes. INSTAP is committed to engaging a variety of audiences by disseminating knowledge through its scholarly publishing program, which produces award-winning monographs that are both academically and popularly acclaimed.

Library of Congress Cataloging-in-Publication Data
Names: Leonard, Albert, Jr. author | Betancourt, Philip P., 1936- author
Title: 5,000 years of fine Cretan wines : glimpses of the history of wine on a beautiful island / Albert Leonard, Jr. and Philip P. Betancourt.
Other titles: Five thousand years of fine Cretan wines
Description: Philadelphia, Pennsylvania : INSTAP Academic Press, 2025. | Includes bibliographical references. |
Summary: "This book discusses the long history of quality wine on Crete. After a historical introduction, the story travels back 5,000 years to eastern Crete where the earliest evidence of a fermented beverage has been discovered on the island. There follows glimpses into the many ensuing centuries including wine from a Minoan shipwreck off the northern coast of the island, the role often played by the god known as Dionysos/Bacchus, and the famous Malvasia wine that many believe was made in Crete and traded widely during the four centuries that Venice occupied the island. Recently, this history has come alive again with the recognition, replanting, and vinification of legacy grape varietals combined with a renewed appreciation by modern Cretan winemakers for the practices used by their forebears."-- Provided by publisher.
Identifiers: LCCN 2025025984 (print) | LCCN 2025025985 (ebook) | ISBN 9781931534437 paperback | ISBN 9781623034481 pdf
Subjects: LCSH: Wine and wine making--Greece--Crete--History | Drinking vessels--Greece--Crete | Crete (Greece)--Antiquities
Classification: LCC TP559.G8 L46 2025 (print) | LCC TP559.G8 (ebook)
LC record available at https://lccn.loc.gov/2025025984
LC ebook record available at https://lccn.loc.gov/2025025985

Front cover: the Mandilari(a) grape. Reprinted from *The Cretan Grapes* (2017, p. 87, Tropi Publications), by M. Stavrakaki and M.N. Stavrakakis; courtesy M. Stavrakaki and M.N. Stavrakakis. See also, Figure 11.23.
Back cover: marble bust of Janus, double herm, Roman copy of a Greek original, Vatican Museums, Museo Chiaramonti, section XIV, no. 17, inv. no. 1395, photo M.-L. Nguyen, CC BY 3.0, https://creativecommons.org/licenses/by/3.0/deed.en, https://commons.wikimedia.org/wiki/User:Jastrow accessed May 16, 2025; photos of authors by T. Curry and S. Ferrence.

Copyright © 2025
INSTAP Academic Press
Philadelphia, Pennsylvania, USA
All rights reserved
It is prohibited without prior authorization to use content of this digital book to train large language models (LLM) or other generative artificial intelligence (AI) technologies or to input into any previously trained AI system.

Dedicated to the Lady of the Vines
Dr. Stavroula Kourakou-Dragona
1928–2024

Table of Contents

List of Figures ix

Preface xix

Acknowledgments xxi

List of Abbreviations xxiii

1. Cretan Wine: A Brief Historical Overview of the Greek Island, *Albert Leonard, Jr.* 1

2. An Introduction to Wine of Ancient Crete, *Philip P. Betancourt* ... 17

3. The Early Bronze Age: An Early Phase in a Long Tradition of Wine in Crete, *Philip P. Betancourt* 21

4. The Middle Bronze Age: Wine Trade in Crete, *Philip P. Betancourt* 25

5. The Late Bronze Age: What the Minoan Tablets Say about Wine, *Philip P. Betancourt* 29

6. Classical Greece: Dionysos/Bacchus, the Greco-Roman God of Wine, *Philip P. Betancourt* 33

7. Cretan Heritage 1: The Wine Glass with a Stem and Base, *Philip P. Betancourt* 39

8. Cretan Heritage 2: The Great Jar Tradition, *Philip P. Betancourt* ... 43

9. Byzantine Wine and the *Geoponika*, *Philip P. Betancourt* 51

10. The Enigmatic Malvasia di Candia Wine, *Albert Leonard, Jr.* 55
11. The Heritage Grapes of Cretan Wine, *Albert Leonard, Jr.* 63
12. The Message on the Bottle, *Albert Leonard, Jr.* 85
13. Glimpses of the Past in the Future of Cretan Wines:
    Resins, Raisins, Pots, and People, *Albert Leonard, Jr.* 95

List of Figures

Frontispiece. Athenian red-figure krater (for mixing wine and water), showing a dinner party with music and wine for the host and his guests. Attributed to the Painter of the Louvre Centauromachy, ca. 450 B.C. ii

Figure 1.1. Sir Arthur Evans seated at Knossos surrounded by a selection of Minoan artifacts excavated under his direction. Oil on canvas portrait by Sir William Blake Richmond, 1907. 2

Figure 1.2. Satellite photo of Crete showing regions and sites in the text. 3

Figure 1.3. Theseus slays the Minotaur as Minos's daughter, Ariadne, watches. Athenian black-figure neck-amphora (jar) attributed to the Edinburgh Painter, ca. 500 B.C. 4

Figure 1.4. A large clay jar called a pithos (height 74.6 cm)—whose inner surface was analyzed by gas chromatography to reveal the ancient contents—was shown to have held wine. It was excavated from the Early Minoan IA site of Aphrodite's Kephali in eastern Crete 4

Figure 1.5. Long storage rooms in the Palace of Minos at Knossos containing large jars that once held wine and other

	agricultural goods in sufficient quantity to last from one harvest to the next, ca. 1430 B.C 4
Figure 1.6.	Satellite photo of the Aegean region showing places mentioned in the text.. 5
Figure 1.7.	Satellite photo of the Mediterranean and Near East showing places mentioned in the text 6
Figure 1.8.	Cretans (Keftiu) shown bearing a range of objects from copper ingots to finished vases sent as tribute to the Egyptian pharaoh. Facsimile painting (tempera and ink on paper) by Nina de Garis Davies after wall painting in the 18th-Dynasty Tomb of Rekhmire (TT 100), Thebes, Egypt, ca. 1479–1425 B.C...................................... 8
Figure 1.9.	A Late Minoan I pyxis (jewelry box) made of wood and elephant ivory and the semi-precious beads it once contained, ca. 1430 B.C. It was excavated on the islet of Mochlos located just off the northeastern coast of Crete .. 8
Figure 1.10.	The hypobranchial gland of the murex sea snail (*Hexaplex trunculus*) was used to produce the coveted reddish-purple dye that became a hallmark product of ancient traders such as the Minoans, Phoenicians, and Romans.......... 9
Figure 1.11.	A stone block from the mid fifth century B.C. showing the Gortyn Law Code in Doric Greek and carved in the *boustrophedon* manner...................................... 10
Figure 1.12.	A silver drachma from Knossos shows the goddess Hera (Zeus's wife) on the obverse and the labyrinth (maze) built by Daedalus to house the Minotaur on the reverse, ca. 350–220 B.C. 10
Figure 1.13.	Watercolor of a late second or early first century B.C. transport amphora fashioned from East Cretan Cream Ware that was excavated at Mochlos in eastern Crete .. 10
Figure 1.14.	Petrographic image of the distinctive clay used in fashioning the East Crete Cream Ware transport amphora.... 11
Figure 1.15.	Byzantine forces under Nikephoros Phokas besiege Chandax 12
Figure 1.16.	The famous bronze horses that would become an iconic symbol of St. Mark's Basilica in Venice were sent there after the siege of Constantinople by Doge Enrico Dandolo and installed on the terrace of the basilica facade in 1254 13
Figure 1.17.	The *koules* (or little fortress) in Herakleion harbor was one part of the *Rocca a Mare* defenses that protected maritime commerce from the Levant all the way home to Venice .. 14

Figure 1.18. An oil on canvas, Portrait of an Old Man, is presumed by many to be a self-portrait of the Cretan-born artist known as El Greco (Domenikos Theotokopoulos), born ca. 1595–1600. 14

Figure 2.1. Installation for pressing grapes to make wine, Gournia, Room D30. 18

Figure 2.2. Athenian red-figure vase by the Cleveland Painter depicting three middle-aged silens making wine by extracting the juice from the grapes while their patron Dionysos watches them busily at work. Ca. 460–450 B.C. 19

Figure 3.1. The small fortified site called Aphrodite's Kephali, located on a hill overlooking the north–south road across eastern Crete, was built about 3000 B.C. 22

Figure 3.2. Consolidating the walls of the small fort at Aphrodite's Kephali. 22

Figure 3.3. The southern end of the fortification wall at Aphrodite's Kephali, overlooking the north–south road leading toward the South Cretan Sea. 22

Figure 3.4. Plan of the site of Aphrodite's Kephali. 23

Figure 3.5. Early Minoan IA pithos no. 77 from Aphrodite's Kephali in eastern Crete 24

Figure 4.1. Divers excavating the Minoan shipwreck near the island of Pseira where a ship sank ca. 1800 B.C. 26

Figure 4.2. Location of the Minoan shipwreck southeast of Pseira island. 26

Figure 4.3. Amphorae made of Mirabello Fabric, probably from Gournia 26

Figure 4.4. Amphorae made of Phyllite Fabric, probably from Mochlos 26

Figure 4.5. View of part of the Minoan town of Gournia, looking south. 27

Figure 5.1. Sealstones of serpentinite from Knossos with Minoan Hieroglyphic signs. 30

Figure 5.2. Clay tablet written in Linear A from the Minoan palace at Kato Zakros accounting quantities of different classes of wine. 31

List of Figures | xi

Figure 5.3.	Comparison of ideograms used for wine in Egyptian Hieroglyphics and the three scripts used in Bronze Age Crete...........31
Figure 6.1.	Dionysos, the god of wine, holding a drinking vessel made from a cow's horn, painted on an Athenian black-figure vase attributed to Lydos. Ca. 550 B.C...........34
Figure 6.2.	A mask of Dionysos painted on an Athenian black-figure krater, a large vessel for mixing water with wine. Ca. 520–510 B.C...........34
Figure 6.3.	Dionysos on a red-figure pelike (jar) by the Geras Painter. The god holds an oversized drinking cup called a kantharos. Ca. 480 B.C...........35
Figure 6.4.	Black-figure amphora featuring the god Dionysos holding his kantharos with two of his followers, a maenad at left and a silen at right. Ca 510 B.C...........35
Figure 6.5.	Ceramic vase of a lounging figure holding a bunch of grapes and sitting beside an amphora, recognizable as a silen by his horse's ears. Fourth century B.C...........36
Figure 6.6.	Marble seats in the Roman Theater of Dionysos, Athens..36
Figure 6.7.	Roman Theater of Dionysos, Athens...........36
Figure 6.8.	Theater at Epidauros, Greece. Fourth century B.C.......37
Figure 6.9.	Theater at Aptera, Crete, with parts of the theater labeled. Fourth century B.C...........37
Figure 6.10.	In Greek mythology, Princess Ariadne was the daughter of King Minos of Knossos and the wife of Dionysos, the god of wine. Athenian red-figure skyphos by the Lewis Painter. Ca. 470 B.C...........37
Figure 7.1.	Large communal goblet made of Pyrgos Ware from the Hagia Photia cemetery, Crete. Early Minoan IB, ca. 2800 B.C...........40
Figure 7.2.	Three goblets of Vasiliki Ware from Vasiliki, Crete. Early Minoan II, ca. 2600 B.C...........40
Figure 7.3.	Goblet from Mycenae, Greece, ca. 1400 B.C...........40
Figure 7.4.	Mycenaean kylix with stylized flower. Late Helladic IIIB:1, ca. 1300–1225 B.C...........41
Figure 7.5.	Athenian skyphos decorated with lines, chevrons, and dots. Ca. 800–750 B.C...........41
Figure 7.6.	Athenian black-figure kylix with siren and "panther." Attributed to Tleson as painter and potter. Ca. 550–540 B.C...41

Figure 7.7.	Gothic chalice with a rounded bowl supported by a stem and base; possibly from Hungary. Silver decorated with gold filigree, enamel, glass, and possibly semi-precious stones. Mid 15th century	42
Figure 7.8.	A violet-tinted glass goblet with additions of gold and small flowers made in a workshop on the island of Murano, Venice, Italy	42
Figure 8.1.	Pithos from Phaistos. Middle Minoan II, ca. 1875–1750 B.C.	44
Figure 8.2.	Excavations of the House of the Metal Merchant at Mochlos uncovered several pithoi. Late Minoan IB, ca. 1625–1470 B.C.	45
Figure 8.3.	Jar from Pseira decorated with a field of spirals. Late Minoan IB, ca. 1625–1470 B.C.	46
Figure 8.4.	Jar from Knossos decorated with an octopus. Late Minoan II–III, ca. 1470–1075 B.C.	46
Figure 8.5.	Traditional turntable used by potters from Thrapsano	47
Figure 8.6.	Potter from Thrapsano building a jar on a traditional turntable set in a trench	47
Figure 8.7.	Potter in 2022 using an electric motor to rotate the turntable	48
Figure 8.8.	Modern jars produced in Thrapsano	48
Figure 8.9.	Garden party at the Tholos Beach Hotel, Kavousi, Crete, with Susan Ferrence and Alessandra Giumlia-Mair admiring a pithari (storage jar) from Thrapsano painted with decoration.	49
Figure 9.1.	The Antioch Chalice, a plain silver goblet enclosed by an elaborate gilded shell with rinceau decoration: grape vines, birds, animals, and seated men. Ca. A.D. 500–550	52
Figure 10.1.	The island of Malvasia as it appeared in the 17th century when it was ruled by the Ottoman Empire	56
Figure 10.2.	The Corinth Canal follows much of the paved (cobblestone) track (*diolkos*) that was begun at the end of the 7th century B.C., connecting the Aegean and Ionian Seas	57
Figure 10.3.	The fearsome winged Lion of St. Mark, depicted on the wall of the Venetian fortress in the harbor of Herakleion and the symbol of the Republic of Venice, stands ever vigilant to protect the island of Crete	58

Figure 10.4.	The Malvasia Bianca di Candia grape	59
Figure 10.5.	A sistrum is a musical percussion instrument associated with the worship of the ancient Egyptian goddess Hathor, and it also is seen in Minoan and Anatolian art: (a) ceramic Middle Minoan examples were excavated at Hagios Charalambos in eastern Crete, ca. 2000 B.C.; (b) Late Minoan I bronze sistrum excavated on the island of Mochlos in eastern Crete; (c) a drawing of a sistrum is featured on the label of Agelakis Winery's *Seistro* wine	60
Figure 10.6.	The top half of the Harvester Vase, excavated in a Late Minoan (ca. 1500 B.C.) villa at Hagia Triada in south-central Crete, shows a sistrum being used to keep the rhythm of a song or chant by a group of agricultural workers	61
Figure 10.7.	The Aeolian Islands off the northeastern coast of Sicily as they are seen from the island of Vulcano with the island of Lipari in the middle, Salina to the left, and Panarea to the right	61
Figure 10.8.	Odysseus (holding trident) is humorously chased across the waves by the North Wind (ΒΟΡΙΑΣ' head at upper right) on a raft consisting of two (wine?) amphorae. Black-figure skyphos (drinking cup) from Thebes, fourth century B.C.	61
Figure 10.9.	The Monemvasia grape	62
Figure 11.1.	The iconic label from a 1960s bottle of *Cretan White Dry Wine* offered by Minos Winery in Peza (a town southeast of Knossos)	64
Figure 11.2.	The Assyrtiko grape	65
Figure 11.3.	By using two Linear B ideograms on its *Assyrtiko* label, Alexakis Winery would have announced to Greek speakers at Knossos over 3,000 years ago that this was a wine that could be enjoyed by both men and women	65
Figure 11.4.	The Athiri grape	66
Figure 11.5.	The Daphni grape	67
Figure 11.6.	Marble sculpture of Apollo and Daphne by Italian Baroque artist Gian Lorenzo Bernini. Ca. 1622–1625	67
Figure 11.7.	The Muscat Spiná(s) grape	68
Figure 11.8.	The Plyto grape	69
Figure 11.9.	The rescued Plyto grape is now enjoying life and thriving in its new home at the Lyrarakis Winery	69

Figure 11.10.	The Thrapsathiri grape 70
Figure 11.11.	The Melissaki grape 71
Figure 11.12.	The Vidiano grape 72
Figure 11.13.	Stressing the relationship between wine and culture, the Endochora Winery uses the image of an early "cupbearer" from the Cycladic Islands on its labels 72
Figure 11.14.	The label for Lyrarakis Winery's *Symbolo* wine commemorates the symbol for wine used over 3,000 years ago on clay tablets in the earliest-known form of the written Greek language, Linear B. 73
Figure 11.15.	The Vilana grape 74
Figure 11.16.	Michalakis Winery pays homage to the island's Minoan past by featuring the image of a rare silver stater (coin) from mid-fifth century B.C. Knossos that depicts a running Minotaur on its *Vin de Crete* Vilana/Vidiano blend.... 74
Figure 11.17.	The Kotsifali grape 75
Figure 11.18.	Endochora Winery takes great pride in its 2021 monovarietal Kotsifali wine with a distinctive label that would stand out on any shelf or table 75
Figure 11.19.	The fresco of the Ladies in Blue from Knossos adorns the label of Boutari's *Kretikos* red wine 76
Figure 11.20.	The Ladikino grape 77
Figure 11.21.	The Liatiko grape 78
Figure 11.22.	Labels from two of Lyrarakis Winery's 2021 monovarietal releases portray what appears to be a pair of very well-dressed women (goddesses?) from Crete's Geometric period (ca. 900–700 B.C.) 79
Figure 11.23.	The Mandilari(a) grape 80
Figure 11.24.	The label on Stylianou's monovarietal Mandilari wine, *Great Mother*, features a woman holding clusters of grapes in imitation of the reptiles that were brandished by the famous Minoan Snake Goddess from Knossos 81
Figure 11.25.	The famous Snake Goddess figurine made of faience was excavated by Arthur Evans in 1903 at the site of Knossos. To many, it has become an icon for the Minoan culture as a whole. Ca. 1600–1500 B.C. 81
Figure 11.26.	The Romeiko grape 82

Figure 12.1. Greek terracotta figurine depicting the cyclops Polyphemus relaxing with Odysseus's gift: a cup of 25-year-old wine made in Ismaros (Thrace) by the legendary winemaker Maron. Perhaps from Boeotia, late 5th to early 4th century B.C. 86

Figure 12.2. A painted sign at the *Ad Cucumas* (At the Jars) wine shop in Herculaneum, Italy, advertised a selection of four different wines, each with its own price. 87

Figure 12.3. The front and back labels for Lyrarakis Winery's 2021 *Vidiano* wine, just as they came from the printer, demonstrate how much important and welcome information can appear on a modern Greek wine label 88

Figure 12.4. For over four centuries, Venice controlled the island of Crete through the administration of four *territoria* (territories). 89

Figure 12.5. The label on Douloufakis Winery's *Dafnios* wine, which uses monovarietal Vidiano grapes, proudly proclaims its PGI (Protected Geographical Indication) status 90

Figure 12.6. Domaine Oikonomou released this 2015 *Sitia* wine as a PDO (Protected Designation of Origin) wine 93

Figure 13.1. *Kretikos* wine label from Boutari Winery, advertising the PGI (Protected Geographical Indication) of the wine .. 96

Figure 13.2. *Minos Palace* wine label 96

Figure 13.3. The antibacterial properties of *Pistacia terebinthus*, also known as the turpentine tree, appear to have been well known to the people of the ancient Mediterranean world 97

Figure 13.4. While the amphora was the preferred method by which to ship wine around the Mediterranean, the wooden barrel was the preferred means on the rivers of Europe, as can be seen in this reproduction of a sandstone funerary monument excavated at Neumagen-Dhron, Germany. The original sculpture memorialized a Roman wine-trader, ca. A.D. 220 98

Figure 13.5. Glass jugs and juglets in the form of barrels were extremely popular in northern and western Europe, especially from the second through the fourth centuries A.D 98

Figure 13.6. A bunch of Savatiano grapes 99

Figure 13.7.   A mosaic "portrait" commemorating Hesiod (*Esio-dus*), the author of *Works and Days*, was excavated at a third century A.D. Roman *domus* (house) in the city of Augusta Treverorum (modern Trier), Germany 100

Figure 13.8 . Another detail of the mosaic from the same Roman house in Trier depicts Octob(er), the eighth month of the Roman calendar, as Dionysos/Bacchus wearing an ivy wreath on his head and carrying the *thyrsus* (magical fennel staff) by which he is identified 101

Figure 13.9.   A statue honoring Columella stands in the Plaza de las Flores (Square of the Flowers) in Cadiz, Spain 102

Figure 13.10. A scene on the interior of an Attic red-figure kylix (drinking cup) from Vulci by the Brygos Painter. Also known as the Iliupersis Cup, it depicts what many interpret as Hecamede pouring a refreshment for Nestor as described by Homer. Ca. 490–480 B.C. 105

Figure 13.11. A late 19th-century winemaker leans against a qvevri (large storage jar) in Kakheti, Republic of Georgia ... 106

Figure 13.12. The Silva-Daskalaki Winery uses indigenous yeasts to ferment the monovarietal wines of their organic Grifos series in large (300-liter) terracotta jars, as shown on their Liatiko wine label 106

Figure 13.13. Domaine Douloufakis produced two wines (a Vidiano and a Muscat Spina) that were fermented in large (250–300 liters) terracotta pitharia with the goal of crafting as natural a wine as possible 107

Preface

In many periods of history, when an artist wishes to show an image to illustrate a pleasant social experience, a glass of wine is an essential ingredient along with music and good friends. The ceramic vessel in the frontispiece, painted by an anonymous vase painter in Athens around the middle of the fifth century B.C., is one example. Wine is not only a beverage. It is also a metaphor for what is pleasant in a quiet and sophisticated way.

This little book provides a series of glimpses of the history of wine over the last 5,000 years on an island in the southern Aegean that is also a metaphor for what is pleasant in life. Crete has a charm all its own, and its countryside and its people, like its wines, should be a metaphor for what is positive in a quiet and sophisticated way.

In his important new book, *The Wines of Crete* (2021), Yiannis Karakasis (Master of Wine, a qualification issued by The Institute of Masters of Wine in the United Kingdom) offered the following advice for promoting Crete and its wines (p. 67): "In a place that has such a wealth of archaeological findings linked to wine, a history that stretches from the Neolithic era to this day, one can weave a beautiful story about this long tradition." The history of wine on Crete is such a long and complex road that it cannot fit within a short book. We prefer to present a series of glimpses along the way, highlighting some interesting episodes. We trust that the pages that follow are a beginning to answering the challenge set forth by Karakasis.

Acknowledgments

In addition to the individuals and institutions mentioned in the text, the authors wish to give special thanks to Maritina Stavrakaki and Manolis Stavrakakis for the kindness they have shown in allowing us to use so many beautiful photographs from their award-winning *The Cretan Grapes* (2017). They make this volume more attractive and more informative.

We also acknowledge the assistance and support of the following people and their institutions who helped to bring this book to fruition either by helping to locate information or clarifying points of confusion: Tim Bell, Tom Brogan, Victor England, Doug Faulmann, Susan Ferrence, Mary Lannin, Eleni Nodarou, Chronis Papanikolopoulos, Elizabeth Shank, Jeff Soles, Dave Stare, and Natalia Vogeikoff-Brogan.

And tremendous gratitude is due to the many wineries that demonstrated legendary Cretan hospitality (ξενία, *xenia*) by welcoming our questions and contributing to our education. The following deserve a special thank you: Agelakis Winery, Alexakis Winery (Stelios Alexakis), Boutari Wines of Crete and the Skalarea Estate (Lenia Boutari, Maria Moschou, Giorgos Michelakis), Domaine Economou (Yiannis Economou), Douloufakis Winery, Endochora Wines (Michalis Tsafarakis), Gavalas Cretan Wines, Iliana Malihin Winery (Iliana Malihin), Lyrarakis Wines (Katerina Lyrarakis), Michalakis Estates Winery, Minos/Miliarakis Winery (Nikos Miliarakis), Pnevmatikakis Winery (Andreas Pnevmatikakis), Silva Daskalakis Wines (Samoli Litsa), Stylianos Winery, and the Titakis Winery.

To all we are truly grateful.

List of Abbreviations

abv.	alcohol by volume	mm	millimeter
acc.	accession number	Mt.	Mount
BIO/bio	organic (often biodynamic) wine	no.	number
		PDO	Protected Designation of Origin
ca.	circa (approximately)		
cm	centimeter	PGI	Protected Geographic Indication
diam.	diameter		
DNA	deoxyribonucleic acid	pl(s).	plate(s)
ECCW	East Cretan Cream Ware	ppm	parts per million
EM	Early Minoan	r.	ruled/reigned
fig(s).	figure(s)	R.S.	residual sugar
g/L	grams per liter	Rust.	*De re rustica* by Columella
GI	Geographic Indication		
INSTAP SCEC	Institute for Aegean Prehistory Study Center for East Crete	suppl.	supplemental
		TT	Theban Tomb
		V.Q.P.R.D.	*Vin de Qualité Produit dans une Région Déterminée* (Wine of Quality Produced in a Determined Region)
km	kilometer		
m	meter		
m.a.s.l.	meters above sea level		
ml	milliliter		

1

Cretan Wine
A Brief Historical Overview of the Greek Island

Albert Leonard, Jr.

The origin of the name Crete (Κρήτη, *Kriti*), by which Homer referred to an island of many cities "in the midst of a wine-dark sea," is uncertain. Biblical writers refer to the island as Caphtor (Deuteronomy 2:23) and trace the origin of the Caphtorites to Egypt (Genesis 10:13–14; 1 Chronicles 1:11–12). Many archaeologists consider Crete to be the place called Keftiu by the Egyptians, or as some variant of the Semitic word Kaptaru used by people elsewhere in the ancient Levant. Today, we know that Crete's ancient population was far more complex than the ancient attributions to a single origin. It was already settled long before recorded history. We refer to Crete's Bronze Age inhabitants as Minoans, a term popularized over a century ago by British archaeologist Sir Arthur Evans (Fig. 1.1) through his excavations at what he called the Palace of Minos near the village of Knossos. Actually, the ancient town of Knossos had been discovered in 1878 by a Cretan antiquarian—appropriately named Minos Kalokairinos—who located it just south of the modern city of Herakleion (Fig. 1.2:16). The original Minos was a mythical king of Crete, said to have been the son of Zeus, who ruled widely by the power of his navy (a *thalassocracy*). His wife Pasiphae had given birth to the dreaded Minotaur—a half bull and half human monster—that was housed in a labyrinth (maze) at Knossos, built by the legendary craftsman Daedalus. The creature was ultimately killed by an Athenian youth named Theseus with the help of Minos's comely daughter Ariadne (Fig. 1.3).

Figure 1.1. Sir Arthur Evans seated at Knossos surrounded by a selection of Minoan artifacts excavated under his direction. Oil on canvas portrait by Sir William Blake Richmond, 1907. Oxford, Ashmolean Museum, acc. WA1907.2; © Ashmolean Museum, University of Oxford; https://collections.ashmolean.org/object/373460, accessed May 16, 2025.

The term Minoan is now used, in one sense or another, to cover many centuries of life across the entire island. This is the period that Aegean prehistorians call the Bronze Age (ca. 3000–1100 B.C.). For Crete, archaeologists further subdivide this period into three major chronological stages: Early, Middle, and Late Minoan. The earliest stage was a period of development. Wine had been a staple of the society for many centuries by this time, and scientific analysis of ceramic jars from the beginning of the Early Minoan period, like the one illustrated in Figure 1.4, stored large quantities of the alcoholic beverage. The Middle Minoan period was a time when great palaces emerged, and Minoan influence spread beyond Crete's borders. The Late Minoan period encompassed both the acme of Minoan society at the beginning of the phase and also the later period when much of the island was ruled by Hellenic people from the mainland of Greece, whom we call the Mycenaeans. By the beginning of the Late Minoan period, wine was one of the necessities of Bronze Age life, because (although the Minoans did not understand the reason) its alcohol killed germs, and those who drank it instead of the often-polluted water did not become ill. Like other needed commodities, it was stored at the palace of Knossos in large magazines (storage rooms) that held enough provisions to last until the next harvest (Fig. 1.5).

The Late Bronze Age was a time of intense international trade and travel when Minoan (and later Mycenaean) Crete was an active participant both as an exporter and an importer. At that time, the peoples of the Mediterranean moved their wares across incredibly great distances: from the Canaanite coast to Sicily, Sardinia, and Spain (Fig. 1.6–1.8). Raw ores and metal ingots, ivory tusks and carved furniture, olives and scented oils, foodstuffs and finished products, grapevine root stock and jars filled with the "gift of Dionysos" were all exchanged. At the upper end of the social scale were the recipients of the *objets d'art* exchanged among the elites of the time (Fig. 1.9). At its lower end, and in the long run more important and lasting, was the widespread use of the skills and talents of the workers who made the products. These were the strengths that would help Crete survive the disastrous migrations of the early 12th century B.C. that accompanied the collapse of civilizations across the eastern Mediterranean. These strengths would be needed as the Bronze Age world transitioned to the Early Iron Age when economic depression took hold in the eastern Mediterranean world.

Figure 1.2. Satellite photo of Crete showing regions and sites mentioned in the text: (1) Alagni, (2) Alatzomouri Pefka, (3) Aphrodite's Kephali, (4) Aptera, (5) Archanes, (6) Azoria, (7) Chryssi, (8) Dafnes, (9) Eleutherna, (10) Eltina, (11) Gortyn, (12) Gournia, (13) Hagia Triada, (14) Hagia Photia, (15) Hagios Nikolaos, (16) Herakleion/Candia, (17) Idaean Cave, (18) Kastelli Kissimos, (19) Knossos, (20) Kommos, (21) Lyttos, (22) Mochlos, (23) Mt. Psiloritis, (24) Peza, (25) Phaistos, (26) Pseira, (27) Rethymnon, (28) Siteia, (29) Spiná, (30) Tylissos, (31) Thrapsano, (32) Zakros. Image J. Papit, H. Sperling; courtesy Google Earth.

Cretan Wine: A Brief Historical Overview of the Greek Island | 3

Figure 1.3. Theseus slays the Minotaur as Minos's daughter, Ariadne, watches. Athenian black-figure neck-amphora (jar) attributed to the Edinburgh Painter, ca. 500 B.C., height 14.6 cm. New York, The Metropolitan Museum of Art, Rogers Fund, 1921, acc. 21.88.92; https://www.metmuseum.org/art/collection/search/251112, accessed May 16, 2025; CC0 1.0, https://creativecommons.org/publicdomain/zero/1.0/.

Figure 1.4. A large clay jar called a pithos (height 74.6 cm)—whose inner surface was analyzed by gas chromatography to reveal the ancient contents—was shown to have held wine. It was excavated from the Early Minoan IA site of Aphrodite's Kephali in eastern Crete. Reprinted from Betancourt 2013, frontispiece; see also Fig. 2.5; photo courtesy P. Betancourt.

Figure 1.5. Long storage rooms in the Palace of Minos at Knossos containing large jars that once held wine and other agricultural goods in sufficient quantity to last from one harvest to the next, ca. 1430 B.C. Evans 1921–1935, IV, suppl. pl. LXI.

Slowly, the archaeological and historical records begin to clear. By the eighth century B.C. life was improving as more tools and weapons of iron slowly took the place of those of copper and bronze. But despite scattered exceptions like the spectacular shields and other eastern material from the Idaean Cave in central Crete (Fig. 1.2:17), the cultural landscape seems to have been more shattered, its once far-reaching vision turned inward upon itself and its island. The power of many of the centers that had prospered when the navy of King Minos ruled the waves had now diminished. Other players had taken the stage, such as the Cretan cities of Gortyn, Lyttos, and Eleutherna, to name a few (Fig. 1.2). And, in uncertain times, pirates were quick to find a wider niche in the island society, although probably not as great a role as the historian Polybius and other ancient authors have suggested.

At some point during this historically murky period, the peoples of the Aegean and those of the Levantine world slowly began to connect

Figure 1.6. Satellite photo of the Aegean region showing places mentioned in the text. Image J. Papit, H. Sperling; courtesy Google Earth.

more closely. When the curtain rises, we find that the Bronze Age Canaanites had become known as the Phoenicians (Φοίνικες, *Phoinikes*), a name derived from the Greek word for reddish purple in reference to the color of the dye (Royal Purple) that these early seafarers extracted from the small murex seashell (*Hexaplex trunculus*). Archaeology has shown that this craft was already practiced during Minoan times at Alatzomouri Pefka (Fig. 1.2:2) in northeastern Crete and on the island of

Cretan Wine: A Brief Historical Overview of the Greek Island | 5

Figure 1.7. Satellite photo of the Mediterranean and Near East showing places mentioned in the text. Image J. Papit, H. Sperling; courtesy Google Earth.

Figure 1.8. Cretans (Keftiu) shown bearing a range of objects from copper ingots to finished vases sent as tribute to the Egyptian pharaoh. Facsimile painting (tempera and ink on paper) by Nina de Garis Davies after wall painting in the 18th-Dynasty Tomb of Rekhmire (TT 100), Thebes, Egypt, ca. 1479–1425 B.C. New York, The Metropolitan Museum of Art, Rogers Fund, 1931, acc. 31.6.42; https://www.metmuseum.org/art/collection/search/544610, accessed January 22, 2025; CC0 1.0, https://creativecommons.org/publicdomain/zero/1.0/.

Figure 1.9. A Late Minoan I pyxis (jewelry box) made of wood and elephant ivory (lid 11 x 14 cm) and the semi-precious beads it once contained, ca. 1430 B.C. It was excavated on the islet of Mochlos located just off the northeastern coast of Crete. In addition to 80 amethyst beads, it contained lapis lazuli that came from Afghanistan. Photo C. Papanikolopoulos; courtesy Mochlos Excavation Project.

Chryssi (Fig. 1.2:7) located in the South Cretan Sea just 15 km south of the modern town of Ierapetra (Fig. 1.10).

At Kommos, fragments of imported Near Eastern storage jars and transport amphorae (vessels with an oval body, narrow cylindrical neck, two handles on the upper part, and a narrow base) suitable for carrying wine appear as early as the 10th century B.C., while other sherds hint at a possible, permanent Phoenician presence there in the ninth century B.C. This combination of imported fragments, in association with local pottery, suggests that a healthy relationship existed between the people of Crete and these maritime merchants. A Phoenician-style shrine built into Temple B in the eighth century B.C. at Kommos further supports the presence of a hybrid community at the site. Toward the end of that period, however, the archaeological evidence for Phoenician contact with Crete begins to fade, possibly due to pressures put on the Levantine homeland by Assyria during the reign of Ashurbanipal (r. 668–627 B.C.). Was this the time for the locals to take the lead? The discovery of a late seventh-century B.C. stone block at ancient Eltina (modern Kounavoi southeast of Herakleion), inscribed with a portion of a legal text, shows that the development of the Greek alphabet was well underway, under Phoenician inspiration.

That the grape was cultivated, and its fruit fermented into wine throughout the Iron Age is obvious from the literary descriptions of

Figure 1.10. The hypobranchial gland of the murex sea snail (*Hexaplex trunculus*) was used to produce the coveted reddish-purple dye that became a hallmark product of ancient traders such as the Minoans, Phoenicians, and Romans. Left: fresh murex (photo courtesy A. Leonard). Right: murex shells excavated from the Late Minoan I island site of Chryssi, located just off the southeastern coast of Crete. Courtesy Archive of the Ephorate of Antiquities of Lasithi. Reprinted from Brogan, et al., 2019, fig. 6.2; photo C. Papanikolopoulos.

winemaking in the works of Homer and Hesiod, both Greek authors of the 8th–7th centuries B.C. On Crete, archaeology offers a glimpse of wine production geared to a local audience. Drinking paraphernalia found in rich burials at several Iron Age sites includes large kraters in which to mix the sweet wine with water, as well as sets of small cups to help share the celebration. The historical record tells us that wine was freely distributed at civic communal meals (*syssitia*), and the excavations at Azoria in eastern Crete suggest an actual building in which such meals would have taken place. But the process appears to be segmented, reflecting the internecine squabbles of an island that at any one time consisted of more than 50 independent, but culturally related, communities, at least in the Iron Age. Cretan winemaking at this time appears to have been in great part a local industry, with distribution confined mainly to the regular pattern of life on the island as hinted in the famous 5th-century B.C. law code at Gortyn. It is written in Doric Greek in the *boustrophedon* (as the ox plows) manner, in which one line was written (and read) from left to right and the next line from right to left (Fig. 1.11). The code offers insight into the daily lives of the Cretans at that time through a series of vignettes such as, to be legal, the process of adoption must include both the exchange of a sacrificial animal and an (unspecified) measure of wine (Column X, lines 33–39).

A broadening of the international trade in wine also was beginning to flourish elsewhere in the eastern Mediterranean. Herodotus tells us that as early as the seventh century B.C., the wines of the town of Mytilene (Fig. 1.6) on the island of Lesvos were exported by Sappho's brother, Charaxus, to members of the cosmopolitan community of Naukratis in far-off Egypt (Fig. 1.7). He even had a girlfriend there, as did many of the Greeks who had served as mercenaries in Egypt since the reign of the pharaoh that they called Psammetichos (r. 664–610 B.C.).

Cretan mercenaries continued to play a role in warfare into the Classical period (most of the 5th–4th centuries B.C.) and later. The skill of Cretan archers was legendary. They had fought in some of the most famous and most significant battles of the day, serving Alexander the Great as well as the generals who competed for power after his death in Babylon in 323 B.C. Perhaps this movement of mercenaries abroad was

Figure 1.11. A stone block from the mid fifth century B.C. showing the Gortyn Law Code in Doric Greek and carved in the *boustrophedon* manner. Courtesy Zde, https://commons.wikimedia.org/wiki/File:Gortyn_code,_ca_450_BC,_Gortys,_145814.jpg, accessed May 18, 2025; CC BY-SA 4.0, https://creativecommons.org/licenses/by-sa/4.0/.

Figure 1.12. A silver drachma from Knossos shows the goddess Hera (Zeus's wife) on the obverse and the labyrinth (maze) built by Daedalus to house the Minotaur on the reverse, ca. 350–220 B.C. Chicago, Art Institute of Chicago, gift of Martin A. Ryerson, acc. 1922.4914, https://www.artic.edu/artworks/5741/drachm-coin-depicting-the-goddess-hera, accessed May 18, 2025; CC0 1.0, https://creativecommons.org/publicdomain/zero/1.0/.

one of the reasons that we begin to see a wider pattern of Cretan pottery and other exports in overseas markets leading up to, and following, the time of Alexander in the fourth century B.C. The Knossian coinage of the day continued to remind the bearer of the glory days when King Minos's navy ruled the commercial waves (Fig. 1.12).

At home in Crete the repertoire of the local potter reflects a turn toward larger shapes that would be more suited to the bulk movement of goods. A case in point is the late second or early first century B.C. transport amphora (Fig. 1.13) from the excavations at Mochlos, fashioned from the distinctive fabric that archaeologists call East Cretan Cream Ware (ECCW) because of the color of its slip. Petrographic analysis of its clays, by Eleni Nodarou at the INSTAP Study Center for East Crete, indicates that this combination of clays is only to be found in the area around Ierapetra in southeastern Crete (Fig. 1.14).

The written record becomes dramatically clearer after Roman general Quintus Caecilius Metellus (Creticus) conquered Crete in 66 B.C., and the island became part of Rome's *Provincia Creta et Cyrenaica*, with the town of Gortyn in central Crete serving as its capital.

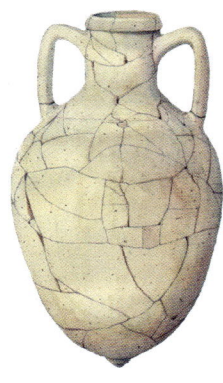

Figure 1.13. Watercolor of a late second or early first century B.C. transport amphora fashioned from East Cretan Cream Ware that was excavated at Mochlos in eastern Crete. Reprinted from Vogeikoff-Brogan 2014, frontispiece; painting D. Faulmann; courtesy Mochlos Excavation Project.

Along with Roman power and structure came the possibility of greater private land ownership. In the ensuing years many of the communities of the hilly hinterland began to move their settlements toward more open areas near the coast. Here they gained easier access to the harbors that would ship their wines and other products to a much wider, international marketplace. Saving a large percentage of the time and cost of transportation from the farmer to the shipper increased Crete's ability to supply competitive, quality wines for the export market. This is demonstrated by the more frequent appearance of its amphorae in archaeological excavations across the Mediterranean, and beyond. In 2017, Italian archaeologists encountered several Cretan amphorae near the House of the Gladiators at Pompeii. Farther afield, the upper portion of a second-century A.D. amphora, with the distinctively pointed handles that archaeologists identify with a potter's workshop on Crete, was found at the Roman legionary fortress at Caerleon in Wales (Fig. 1.7). It bore the painted address LEG II AVG (Second Legion, Augustus) on its neck, and its contents were described on its shoulder as the "best quality raisin wine," echoing the sentiments of Pliny the Elder that Cretan wine (named *passum*) was the finest wine of its class (Pliny, *Natural History* 14.11).

Words of praise such as this caused Cretan *passum* (wine from semi-dried grapes) to be widely acclaimed, traded, and appreciated in the far markets of the Mediterranean. In fact, one of the very few Roman authors who is thought to downplay the excellence of Cretan wine was Martial in his Epigram 106 (Book XIII; see below), written sometime between the years A.D. 86 and 103. It is one of hundreds of pithy epigrams (two-line couplets) intended to accompany, and cleverly introduce or explain, a selection of gifts that Martial suggested one might offer to a guest. His appraisal of a gift of wine from a vineyard at Knossos has been translated variously as "the honeyed wine of the poor man" or "a poor man's mead," and both statements have been considered to be negative. But, on a closer reading, when taken together with Epigram 108 (Book XIII; see below), Martial's reference can be understood in a very positive light.

The Knossian *passum* that Martial describes was a late harvest wine made from grapes that had subsequently been partially raisined in the manner described by Homer and Hesiod in the eighth century B.C. (see below, pp. 99–100). Falernian wine, on the other hand, from vineyards on Monte Masaccio north of Naples, received much of its desired sweetness through the addition of honey. It was said to be the wine of the gods, worthy enough to be featured at a banquet held by Julius Caesar to celebrate his conquests in Spain in 60 B.C. Pliny and other ancient authors lauded Falernian wine as one of the finest ever made. It seems that

Martial's Epigram 106. Raisined Wine (Passum)

"The vineyard of Knossos in Crete where Minos reigned, produced this (wine) for you. It is the honeyed wine of the poor man."

Martial's Epigram 108. Honeyed Wine (Mulsum)

"You, Attic honey, thicken the nectarous Falernian wine. It is only proper that such a beverage should be mixed by a Ganymede."

Figure 1.14. Petrographic image (width of field 4.4 mm) of the distinctive clay used in fashioning the East Crete Cream Ware transport amphora (see also Fig. 1.13). It proves that the pot was crafted in the southern part of the island. Reprinted from Boileau and Whitbread 2014, pl. 15; courtesy Mochlos Excavation Project.

the comparison Martial is making in his epigrams is not between Cretan *passum* and another *passum* made in a similar manner elsewhere, but rather it is a comparison between two sweet wines: *passum* and *mulsum*. The former was a naturally sweet (grapes only) product, while the latter gained much of its sweetness from the addition of the best (and most expensive) honey. Martial is reminding us that Cretan wine was as good (i.e., as sweet) as you can get *naturally*, without the added expense of the honey!

Not a great deal is known about daily life on the island when the Roman Empire split after the death of Theodosius I in A.D. 395, other than the fact that it was absorbed slowly into the affairs of Constantinople and the Eastern Roman Empire, also called the Byzantine Empire. Wine was still exchanged for personal and communal use at the local level, but the halcyon days of far-flung markets had ended. Orthodox and Catholic communities would still need sacramental wine for their Eucharist services, with the only requirement being that it followed tradition, and that it was made from pure (clean) grapes that had been fermented to a sweet wine.

With the island's capital still at Gortyn in central Crete (Fig. 1.2:11), life on the island continued along the same path well into the 820s when a group of wandering Saracens—people who originally came from Arabia and who were then evicted from Iberian Al-Andalus (Fig. 1.7)—captured the island. Their leader, Abu Hafs, founded the Arab Emirate of Crete. The new capital of the island, *Rabaḍ al-Khandaq* (Castle/Fortress of the Moat), often shortened to Chandax or Chandakas, was built where modern Herakleion stands today (Fig. 1.2:16). Through the years it has been referred to as Το Μέγαλο Κάστρο (*To Megalo Kastro*, The Great Castle), the main Muslim castle/fortress on the island.

After several unsuccessful Byzantine attempts to recapture the island of Crete, the future emperor Nicephorus II Phocas (r. A.D. 963–969) laid

Figure 1.15. Byzantine forces under Nikephoros Phokas besiege Chandax: ships in water at left with soldiers and tents in center, located outside the city represented at right. Reprinted from J. Skylitzes, *Madrid Skylitzes*, 12th century, a version of the *Synopsis of Histories*, in the Biblioteca Nacional de España, Madrid, illustrator unknown, https://en.wikipedia.org/wiki/Siege_of_Chandax#/media/File:Byzantines_under_Nikephoros_Phokas_besiege_Chandax.png, public domain, accessed May 18, 2025.

siege to Chandax during the winter of 960–961, and this time Byzantine efforts were successful. The perils of the time were chronicled by an unknown 12th-century painter in the richly illustrated *Synopsis of Histories*, written by John Skylitzes, an 11th-century Byzantine historian (Fig. 1.15). The Muslim defenders were routed by early March (A.D. 961) and, once again, Constantinople was able to connect with its markets in North Africa without fear of being harassed by Saracen corsairs safely harbored in ports along the northern coast of Crete. The name of the Muslim capital Chandax was Latinized to Candia, and the island remained in Byzantine hands until the turmoil of the early 13th century.

Little is known of Cretan winemaking during these two periods of Byzantine rule, or the Muslim interlude between them, but the 10th century encyclopedia *Geoponika* (see Ch. 9) praises the sweet wines of Bithynia (a region in northwestern Anatolia), and its author—possibly Constantine VII Porphyrogenites himself—offers advice on how to obtain the grape's maximum sweetness through a late harvest (Book XVIII). These wines, made and distributed on the local level, would have been in demand by both the Christian and Jewish communities for their religious observations. The growing accumulation of vinicultural knowledge in the hands of the church at this time should also be noted.

The 13th century opened with the Fourth Crusade (1202–1204). It was originally called by Pope Innocent II to liberate Muslim-controlled Jerusalem and the Holy Land, but it was diverted to Constantinople through the shrewdness and deceit of Venetian Doge Enrico Dandolo. On the morning of April 12th, 1204, the city that had been founded a millennium earlier by Constantine the Great was sacked. The following year, as part of the *spolia* (spoils of war), Crete was sold to the Republic of Venice, and the treasures of Constantinople were hustled off to the city traditionally known as La Serenissima (Fig. 1.16).

Under the Venetians Crete became known as il Ducato di Candia (the Kingdom of Candia/Crete), an entity that would later be expanded to include the neighboring island of Kythera, and, eventually, the island of

Figure 1.16. The famous bronze horses that would become an iconic symbol of St. Mark's Basilica in Venice were sent there after the siege of Constantinople by Doge Enrico Dandolo and installed on the terrace of the basilica facade in 1254. For their protection they were replaced by copies in the 1980s, and the originals are installed inside the church. Ca. second or third century A.D. Photo M. Landon, https://commons.wikimedia.org/wiki/Category:Horses_of_Saint_Mark_(Venice)#/media/File:Venice_S_Marco_horses_02.jpg, accessed May 19, 2025; CC BY 4.0, https://creativecommons.org/licenses/by/4.0/.

Tinos far to the north (Fig. 1.6). So, in some cases, the word Candia could refer to the city itself, the entire island of Crete, or all three islands as a unit. With funds fueled by a Mediterranean-wide market for the popular *Malvasia* wine (see Ch. 10), a heavily fortified Crete (Fig. 1.17) entered a long period of economic success. For over four centuries both painting and literature flourished under Venetian hegemony (see especially Manolioudis, Further Reading). The Cretan School of icon painting matured into the Cretan Renaissance when ample additions of western influence were added to its Byzantine base. Herakleion-born artist Doménikos Theotokópoulos (perhaps better known to the modern world as El Greco) went on to influence the Spanish Renaissance through his unique blending of artistic styles (Fig. 1.18). Cretan literature reached an apex with the tragedy *Erofíli* by Georgios Chortatzis; Vikentios Kornaros's *Erotokritos*, a long romantic poem in the Cretan dialect of the early 17th century, was resurrected for modern ears by the lyra of Nikos Xilouris (*Psaroniko*). But all of this ended when Venice was forced to surrender Crete to the Ottomans.

The ride had been long, but it hadn't been free. Cretan agriculture had been watched closely and governed by the Venetian bureaucracy. Cretan interests had been overridden constantly to fill the stomachs of Venice, and to generate the profits that would fill the purses of its tax-paying merchants through the sale of Cretan wines. In Shakespeare's

Figure 1.17. The *koules* (or little fortress) in Herakleion harbor was one part of the *Rocca a Mare* defenses that protected maritime commerce from the Levant all the way home to Venice. Completed in 1540, the *koules* protected Candia for over a century until it was captured by the Ottomans in 1669. Photo courtesy A. Leonard.

Figure 1.18. An oil on canvas, Portrait of an Old Man, is presumed by many to be a self-portrait of the Cretan-born artist known as El Greco (Domenikos Theotokopoulos), born ca. 1595–1600. New York, The Metropolitan Museum of Art, Joseph Pulitzer Bequest, 1924, acc. 24.197.1, https://www.metmuseum.org/art/collection/search/436574, accessed May 19, 2025; CC0 1.0, https://creativecommons.org/publicdomain/zero/1.0/.

The Merchant of Venice, Portia (disguised as a lawyer) confronts Shylock near the Rialto Bridge and tells him that mercy "blesses him who gives, and him who takes" (Act 4, Scene 1). Could she really have been thinking of the Venetian regulations on the transshipment of wine that taxed that commodity once when it came into La Serenissima, and once again when it left? It is a small wonder that there are so many street names in Venice that are formed by a variant of Calle di Malvasia.

The Ottoman invasion of Crete began in 1645 when ca. 40,000 Muslim troops landed on the western end of the island. For more than two decades Candia persevered, holding out during the longest siege in the history of Europe. But on September 27th, 1669, the Ottomans breached the Great Fortress and killed the golden goose of the waning Venetian Empire. For the next 200 years Ottoman Kandiya would mostly emphasize its table grapes (sultanas). There was little place for the alcohol of Malvasia in Islam and, except for a few rulers like Sultan Selim "the Sot" (Sarhos Selim who ruled from 1566 to 1574), there was little place for it in the Ottoman Province (Eyalet) of Crete. Wine production was frowned upon—or worse. Fortunately, the age-old techniques of producing thick, viscous, sweet wines by raisining the grapes survived in cottages across the island, but output shrank. There was no market to drive the need and desire for a greater supply, and, as had happened so often in the past, Crete turned inward upon itself again.

Absent the long days of sunshine enjoyed by the Mediterranean world, Europe to the north would increasingly be forced to drink thinner, dryer, weaker wines compared to wines today. Only the faintest memory would remain of the sweet, powerful wines that had been gathered at the Peloponnesian port of Monemvasia (Fig. 1.7) from sunny, southern islands such as Crete and Cyprus. No longer would these wines be transshipped safely along the Strato da Màr to Venice, from where they could be traded on to the rest of Europe. Slowly these memories would continue to fade until they reached the point when any sweet wine could dimly claim to be the true Malvasia di Candia.

But the memories were not dead; they were merely dormant, and like the vines of winter, they were destined to blossom again.

Further Reading

Betancourt, P.P. 2013. *Aphrodite's Kephali: An Early Minoan I Defensive Site in Eastern Crete* (*Prehistory Monographs* 41), INSTAP Academic Press.

Boileau, M.-C., and I. Whitbread. 2014. "Petrographic Analysis of Local and Imported Transport Amphorae from Knossos, Mochlos, and Myrtos Pyrgos," in *Mochlos III: The Late Hellenistic Settlement. The Beam-Press Complex* (*Prehistory Monographs* 48), by N. Vogeikoff-Brogan, INSTAP Academic Press, pp. 79–102.

Bostock, J., and H.T. Riley, trans. 1855. *The Natural History of Pliny*, Henry G. Bohn.

Brogan, T.M., D. Mylona, V. Apostolakou, P.P. Betancourt, and C. Sofianou. 2019. "A Bronze Age Fishing Village on Chryssi," in *Exploring a* Terra Incognita *on Crete: Recent Research on Bronze Age Habitation in the Southern Ierapetra Isthmus*, K. Chalikias and E. Oddo, eds., INSTAP Academic Press, pp. 97–109.

Chalikias, K., and E. Oddo, eds. 2019. *Exploring a* Terra Incognita *on Crete: Recent Research on Bronze Age Habitation in the Southern Ierapetra Isthmus*, INSTAP Academic Press.

Dalby, A. 2011. *Geoponika: Farm Work. A Modern Translation of the Roman and Byzantine Farming Handbook*, Prospect Books.

Evans, A.J. 1921–1935. *The Palace of Minos at Knossos* I–IV, Macmillan and Co., Limited.

Fleming, S.J. 2001. *Vinum: The Story of Roman Wine*, Art Flair.

Hood, S. 1971. *The Minoans: The Story of Bronze Age Crete*, Thames and Hudson.

Leonard, A. 1995. "'Canaanite Jars' and the Late Bronze Age Aegeo-Levantine Wine Trade," in *The Origins and Ancient History of Wine*, P. McGovern, S.J. Fleming, and S.H. Katz, eds., Gordon and Breach Publishers, pp. 233–255.

———. 2020. *Mediterranean Wines of Place: A Celebration of Heritage Grapes*, Lockwood Press.

Manolioudis, S.M. No date. *Wine Routes in the Cultural Landscapes and Malvasia of Crete*, Herakleion.

McGovern, P.E. 2003. *Ancient Wine: The Search for the Origins of Viticulture (Princeton Science Library)*, Princeton University Press.

McGovern, P.E., S.J. Fleming, and S.H. Katz, eds. 1995. *The Origins and Ancient History of Wine (Food and Nutrition in History and Anthropology* 11), Gordon and Breach Publishers.

Morris, J. 1980. *The Venetian Empire: A Sea Voyage*, 1st ed., Faber.

Sanders, I.F. 1982. *Roman Crete: An Archaeological Survey and Gazetteer of Late Hellenistic, Roman and Early Byzantine Crete (Classical Studies)*, Aris and Phillips.

Shackleton Bailey, D.R., ed. and trans. 1993. *Martial: Epigrams*, vol. III: books 11–14 (*Loeb Classical Library* 480), Harvard University Press.

Skylitzes, J. 2010. *A Synopsis of Byzantine History, 811–1057*, trans. J. Wortley, Cambridge University Press.

Vogeikoff-Brogan, N. 2014. *Mochlos III: The Late Hellenistic Settlement. The Beam-Press Complex (Prehistory Monographs* 48), INSTAP Academic Press.

Wright, J.C., ed. 2004. *The Mycenaean Feast*, special issue, *Hesperia* 73 (2), American School of Classical Studies at Athens.

2

An Introduction to Wine of Ancient Crete

Philip P. Betancourt

In addition to the written sources, the popularity of wine in ancient Crete also is documented by archaeological excavations and other types of evidence. Both amphorae for shipping wine and installations for pressing grapes have been found at several locations. Ancient texts furnish additional information, and the analysis of pottery for microscopic traces of the wines also has contributed useful data.

Wine is not difficult to make if one has good grapes and knows the process. Some of the basic information about Minoan wines and how they were made comes from the analysis of the ancient contents of ceramic vases, based on minute traces of what was absorbed into the pottery fabric. The analytical method used for this work, called gas chromatography, reveals the chemistry of the ancient traces, and their composition provides clues to how the contents of the vessels were produced. Analysis has shown that most Minoan wines were red, which means that the fermentation first occurred while the skins were still on the grapes, suggesting a two-stage process. It also has revealed that some of the ancient wines contained wood resin. Resinated wine, called retsina, is still a popular variety of wine in Crete today. The resin must mean that the aging process occurred in containers that were waterproofed with this material.

The soils of Crete are very favorable for grape growing. The Mediterranean climate, with warm winters and dry summers, is also very conducive to growing this plant. At present, the island furnishes about 15% of the grapes grown in Greece. This is a substantial percentage, and it is a testament to the climate and the topography as well as to the soils of the south Aegean island. The modern situation is not new; Crete's terrain was also suitable for viticulture in other periods of its history.

Although no direct information on the Minoan cultivation and harvesting of grapes survives from archaeology, more recent practices, plus information Minoan accountants kept of various classes of wine, suggests that many different types of wines existed. Competition between winemakers had already succeeded in teaching the growers that harvesting is best when the grapes are at their peak of ripeness, because, before they are ripe, the grapes do not have enough sugar content, and unlike some fruits (like bananas), grapes stop ripening when they are picked.

Grapes have natural yeasts, so they can ferment into alcohol without the necessity of adding this ingredient. The color of red wines is caused by fermentation while the skins are still on. After they were picked, the grapes would have been allowed to start fermenting; then they would have been pressed after the fermentation had progressed sufficiently. The Cretan Bronze Age pressing system was a relatively simple affair. It consisted of a basin on a platform or bench and a lower container to collect the grape juice. An example from Gournia is typical (Fig. 2.1). Both the large basin for holding the grapes while they were trod and the container for the juice were made of clay.

This simple arrangement must have worked in a satisfactory manner because it survived into the Classical period of Greece. The use of an installation of this type is illustrated by a lively red-figure vase created by an Athenian vase-painter known as the Cleveland Painter (Fig. 2.2). The artist was active just before the middle of the fifth century B.C. The image shows wine being made by the followers of Dionysos, the mythological god of wine. The artist imagined the half-human and half-horse silens (note their horse tails) as middle-aged wine makers. One fellow brings an amphora, perhaps with the partly fermented grapes, a second one steps on the grapes in the basin to extract the juice, while a third figure gestures and provides free but not necessarily helpful advice. The whole operation is observed by Dionysos, who stands quietly at the right of the scene.

Figure 2.1. Installation for pressing grapes to make wine, Gournia, Room D30. Reprinted from Hawes et al. 2014, 27, fig. 11.

Figure 2.2. Athenian red-figure vase by the Cleveland Painter depicting three middle-aged silens making wine by extracting the juice from the grapes while their patron Dionysos (far right) watches them busily at work. Ca. 460–450 B.C., height 32.5 cm. New York, The Metropolitan Museum of Art, Rogers Fund, 1941, acc. 41.162.10, https://www.metmuseum.org/art/collection/search/254177, accessed May 19, 2025; CC0 1.0, https://creativecommons.org/publicdomain/zero/1.0/.

The image on the vase illustrates a necessary step in making wine, the removal of the liquid from the skins and seeds. It is possible that after this process was completed the wine would then be poured through a finely woven piece of cloth to remove unwanted sediment or the occasional seed. Purifying the wine by this method or some other way would have been necessary.

Little evidence survives for the next step in ancient Cretan wine making, which is the continuation of the fermentation and the aging process (called malolactic fermentation) when malic acid is converted into lactic acid. It also would have been possible to introduce special flavors to the wine at this stage by inserting additional ingredients. The aging process would require an environment whose temperature did not fluctuate, like the storage magazines with rows of pithoi (very large jars with wide mouths) in the low levels of the Palace of Minos at Knossos (Fig. 1.5).

While it is possible that the malolactic fermentation continued naturally, one cannot rule out that it was induced or improved by the introduction of the proper bacteria by adding them from wine made the previous year. Keeping the wine in a stable and cool environment is important during the aging process because unwanted bacterial growth of the wrong type can induce spoilage. Fluctuations in temperature during this process are not desirable. One can note that Crete has thousands of caves that would be useful for this stage, and that the excavation of East Cretan domestic contexts in towns like Pseira has shown that many Minoan houses had a special storage room at the center of the home that had no doors and could be entered only through an opening in the ceiling. Such a room in a stone building with an earthen floor would give the space an especially cool and stable temperature, well suited for aging wines.

After the aging process, the wines would be shipped to their chosen destinations. The amphora, a transport container with two handles on the upper part and a narrow base, was the most common shipping vessel, especially for travel by sea. The shipwreck at Pseira (see Ch. 4) had a cargo with over 50 amphorae of this type. The same excavation has shown that similar vessels with only one spout instead of a pair also were used for shipping. For storage in both palaces and private dwellings, bulk storage of large amounts was both in small ceramic vessels and in the large jars that are called pithoi.

Further Reading

Curtis, R.I. 2001. *Ancient Food Technology (Technology and Change in History 5)*, Brill.

Hawes, H.B., B.E. Williams, R.B. Seager, and E.H. Hall. 2014. *Gournia, Vasiliki and Other Prehistoric Sites on the Isthmus of Hierapetra, Crete: Excavations of the Wells-Houston-Cramp Expeditions 1901, 1903, 1904*, 2nd ed., INSTAP Academic Press.

3

The Early Bronze Age
An Early Phase in a Long Tradition of Wine in Crete

Philip P. Betancourt

Less is known about Early Bronze Age wine making in Crete compared to later periods. The historical phase, called Early Minoan on the island, is divided into three periods (abbreviated EM I, EM II, and EM III). It lasted about 1,000 years, from just before 3000 B.C. to about 1900 B.C. An example of an archaeological site from this early period with evidence for wine is the small hilltop fort of Aphrodite's Kephali (Fig. 1.2:3). By the time it was built, wine making was already an established tradition in Crete.

Aphrodite's Kephali is a unique ancient archaeological site. It is a small hilltop fort in eastern Crete, perched on the crest of a hill or *kephali* (the local Greek word for a small hilltop) overlooking the road leading north from the south side of the island (Figs. 3.1, 3.2). It stands well above the flat valley that is the only easy route north and south across this mostly mountainous isle. Some of the earliest evidence for wine stored in large jars called pithoi comes from this little walled fort, and the important, actually crucial, role that wine played during this very early time period can best be understood by beginning with the nature of the fort itself.

Aphrodite's Kephali was excavated in 1996 under the direction of Theodore Eliopoulos for the local branch of the Greek archaeological service (*ephoreia*). The excavation uncovered much of the architecture, and a second season of work directed in 2003 by Vili Apostolakou, the director of the East Cretan Ephorate, uncovered the rest. It was obvious from the pottery fragments that this was a significant archaeological site from the very beginning of the Minoan period. Because it was the first site of

Figure 3.1. The small fortified site called Aphrodite's Kephali, located on a hill overlooking the north–south road across eastern Crete, was built about 3000 B.C. Photo P. Betancourt.

Figure 3.2. Consolidating the walls of the small fort at Aphrodite's Kephali. Photo P. Betancourt.

Figure 3.3. The southern end of the fortification wall at Aphrodite's Kephali, overlooking the north–south road leading toward the South Cretan Sea. Photo P. Betancourt.

this type to be excavated, it filled an important gap in our knowledge about the beginnings of the Minoan civilization, the society that went on to create the brilliant palaces of Bronze Age Crete.

The Minoan society flourished for 2000 years on the Greek island of Crete. At its height around 1500 B.C., it was the most advanced civilization of Europe with written documents, large palaces, brilliant art works including fine fresco paintings, and a maritime trading system that linked the early civilizations of western Asia and Egypt with the central and western Mediterranean regions and the rest of Europe (Fig. 1.7). Its main city, Knossos, was the largest city in Europe (Fig. 1.6). Aphrodite's Kephali provides a glimpse of the civilization at its very beginning, a time when new settlers from the northeast were bringing fresh ideas into this part of the Mediterranean basin, hundreds of years before those ideas would result in the first civilization on European soil.

The excavation and study of the site, which was published in 2013, yielded very surprising results in several different areas. The building was designed for defense during a very dangerous period in Cretan history at the end of the Stone Age and the beginning of the Bronze Age, just after 3000 B.C. It was obvious that this was a defensive construction built because real danger threatened the local residents. It faced the south, where a long valley led directly to the seacoast on the southern side of Crete (Fig. 3.3). The fort itself was well laid out (Fig. 3.4). It had a large empty space at the center and covered rooms (called casemates) within the fortification wall. This construction system provided a flat area on the top of the protective wall so defenders could move to wherever they were needed during an attack. A large solid platform at the south, either a lookout point or perhaps a base for a tower, offered a view of the road as far as it extended to the south coast 7 km away. A fire area in the courtyard was large enough to build bonfires to summon people from the local area to come to the shelter of the fort if danger threatened, and an open courtyard provided room for them and some of their animals. A cave inside the walls at the southeast would have been a place for storage and perhaps a source of water. Most importantly, the fort had been well stocked with provisions to last some time, stored in both large and small clay containers.

The analysis by gas chromatography of pottery fragments to discover vessel contents provided much new information. Many of the clay vessels held red wine, and a few contained olive oil. Surprisingly, some of the wines had a resin flavor like the modern Cretan wine called retsina. Evidently the wines had been stored in containers that added the flavor to the beverage. One of the large pithoi analyzed by this method contained red wine (Figs. 1.4, 3.5) and could have held 150 liters of liquid if it was filled to the bottom of the collar and 165 liters if filled to the brim. At the end of its history, it had been broken and mended for reuse, so evidently it was a wine container before it cracked. The fact that it was mended indicates the high value of these large jars, the production of which was a major technological achievement in this early period coming just after the Neolithic era. The kiln for firing them had to maintain an even temperature inside the firing chamber without any colder spots to keep the vessels from breaking due to unwanted uneven temperatures in different places. It represents the work of a real master kiln builder.

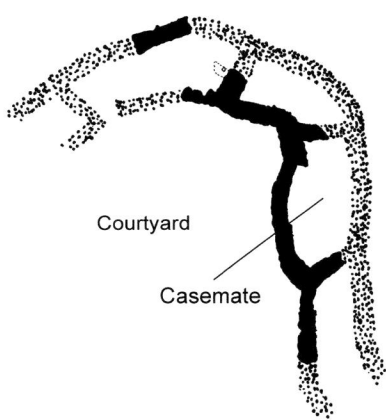

The fact that the presence of wine was detected in several of the clay containers, both large and small, and that they were in long-term storage in a fort that would have been used as a communal shelter in times of danger, suggests an interesting conclusion about the wine. Probably some of the vessels held food that was stored using wine as a preservative. Refugees seeking shelter from pirates or other hostile forces would need the long-term storage of foods, not just wine. Unfortunately, foods preserved in the alcoholic beverage are unlikely to yield a residue from soaking into the clay

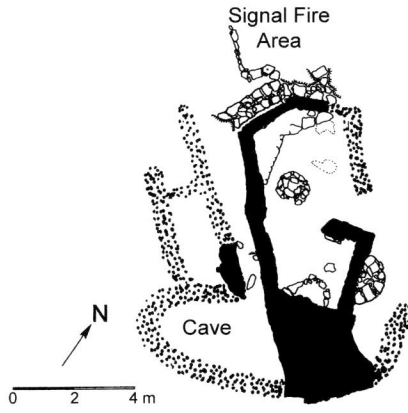

Figure 3.4. Plan of the site of Aphrodite's Kephali. The large base at the south probably supported a tower facing toward the coast. A signal fire area is just north of the small building at the south of the courtyard. The partly preserved fortification wall includes compartments called casemates. Drawing P. Betancourt.

vessel wall, but it is a good speculation that jars must have held foods of various types preserved in the wine.

In a society without any refrigeration and that knew nothing about microbes, the fact that items stored in wine did not spoil like those left in the air would be important knowledge. It would make wine an essential part of society, not just as a beverage by itself, but also as a way to give foods a pleasant flavor and help them last until the next growing season. Jars of fruits, vegetables, and other farm produce preserved in wines would become a valuable trade item and a class of wealth that could be stored, traded, inherited, or even stolen. The large amount of vessels of this type amassed at Aphrodite's Kephali needed walled protection with armed defenders to guard it if the sanctuary was to survive from hostile seafarers who might land on Crete's unprotected shores. Like Aphrodite's Kephali, almost all of the settlements from this period in southern Crete were well inland, on high hills, and surrounded by defensible walls. The Minoans were already actively trading goods within their large island, but the seacoasts were not yet safe during this early period of Cretan history.

Further Reading

Betancourt, P.P. 2013. *Aphrodite's Kephali: An Early Minoan I Defensive Site in Eastern Crete* (*Prehistory Monographs* 41), INSTAP Academic Press.

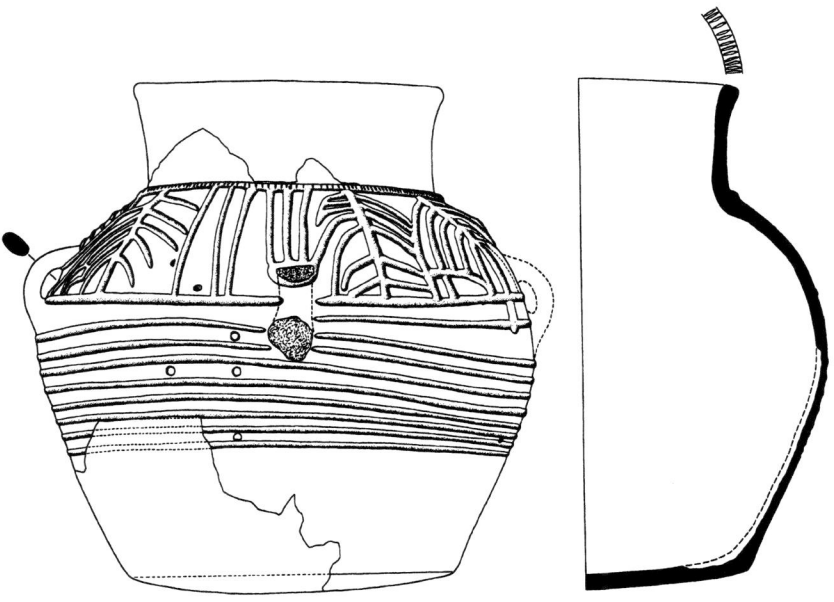

Figure 3.5. Early Minoan IA pithos no. 77 (height 74.6 cm) from Aphrodite's Kephali in eastern Crete. Reprinted from Betancourt 2013, fig. 8.6; drawing D. Faulmann.

4

The Middle Bronze Age
Wine Trade in Crete

Philip P. Betancourt

Crete had many independent regions during the Middle Minoan period (2000–1500 B.C.) because the expansion of the city of Knossos to a dominant position on the island was still in the future. During the Middle Minoan period, the population grew and settlements expanded. Many of them improved their economic situation by increasing their production of agricultural products to get a larger surplus for trade. Because Crete needed metals and other items that were not available locally on the island, these products needed to be imported from overseas. One of the local Cretan items that could help the new trade needs was wine. The excavation of a small Minoan shipwreck loaded with amphorae and other goods helps document the picture (Fig. 4.1).

The shipwreck was found near the small offshore island of Pseira (Fig. 1.2:26) in the Gulf of Mirabello in northeastern Crete (Fig. 4.2). Pseira had been settled since the Neolithic period, and like Crete itself, its residents spoke the Minoan language. The ship's cargo consisted of clay transport vessels for various commodities. It sank during the Middle Bronze Age (ca. 1700/1600 B.C.). The underwater site has been excavated admirably and published by Elpida Hadjidaki-Marder, and as the first Minoan shipwreck to be discovered, it provides a wealth of new information about early Cretan shipping. Most of the cargo consisted of amphorae and other shapes that held liquids, suggesting that it was either the ship of a wine merchant or of a trader who was particularly interested in wine.

Figure 4.1. Divers excavating the Minoan shipwreck near the island of Pseira where a ship sank ca. 1800 B.C. Photo courtesy E. Hadjidaki-Marder.

Figure 4.2. Location of the Minoan shipwreck southeast of Pseira island (note the ship of the underwater excavation team floating above the wreck), looking east toward the coast of Crete. Photo S. Ferrence.

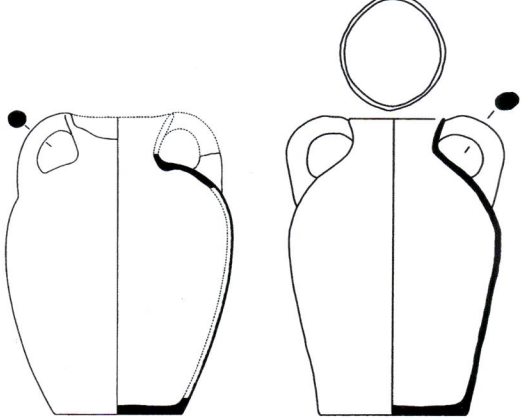

Figure 4.3. Amphorae made of Mirabello Fabric, probably from Gournia, height ca. 37 cm. Drawings L. Bonga.

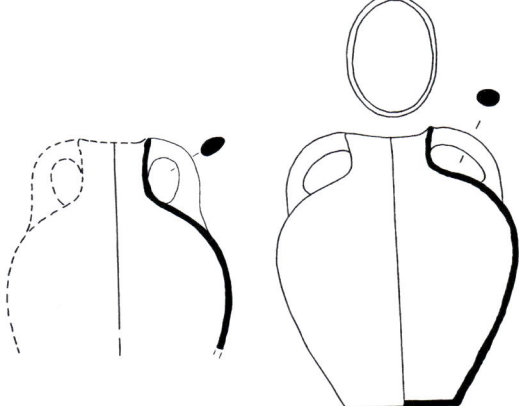

Figure 4.4. Amphorae made of Phyllite Fabric, probably from Mochlos. Right example: height ca. 31 cm. Drawings L. Bonga.

Although none of the wood of the ship survived, the arrangement of the ceramic vessels and other objects on the seabed was sufficient to recognize items owned by the crew as well as the contents of the cargo. Amphorae of two main types were present, with some of the examples originating from Gournia (Figs. 1.2:12; 4.3) along with a class from farther east along the coast of Crete, either Mochlos (Fig. 1.2:22) or a nearby site (Fig. 4.4). The transport amphora, a large vessel with two handles near the rim and a narrow lower part designed to make the vessel stay in place without rolling around in the hold of a ship, was used most often to ship wine. The large cargo of amphorae suggests the possibility that this ship may have been a wine merchant. It was certainly carrying a large cargo of the typical shape for this product.

Many of the transport vessels came from Gournia, a seaport that was the largest settlement in this region of Crete. It is located farther west of the shipwreck's location along the same coastline (Fig. 4.5). The residents had recently constructed a small palace for their rulers, and it is possible that the ship had some relationship to the large town because it had loaded much of its cargo at that location. It was traveling along the east–west trade route on the northern coast of Crete when it had the misfortune to sink.

The Pseira shipwreck is a good example of what was probably a common sight at the busy ports of Crete. It was one of the small local ships that carried whatever needed to be transported from place to place along the coast of the large Aegean island because the mountainous interior of Crete made overland travel much slower, especially if heavy goods needed to be carried. The shipwreck documents the fact that wine was already being produced in quantities that were larger than what was consumed at home, and that its quality was high enough to make it a valuable commodity for trade.

Further Reading

Hadjidaki-Marder, E. 2021. *The Minoan Shipwreck at Pseira, Crete* (*Prehistory Monographs* 65), INSTAP Academic Press.

Figure 4.5. View of part of the Minoan town of Gournia looking south. Photo P. Betancourt.

5

The Late Bronze Age
What the Minoan Tablets Say about Wine

Philip P. Betancourt

Although many of the details are not known, some important evidence about wine comes from the clay tablets of Minoan Crete, reflecting its gradually growing importance in the island's economy. The evidence for Bronze Age wine making from the clay tablets includes several pieces of information that are not available in other ways. Writing was used for many purposes in ancient Crete. It developed locally on the island, and it first became popular as a way of adding short messages to seals that were used to sign documents. All of the Cretan scripts used tiny pictures to stand for sounds in the spoken language, but after the first experiments, the images became so simplified that we can seldom recognize exactly what their origins were.

The writing systems changed twice during the approximately 2000 years of the Minoan Bronze Age. No written records of any kind survive from the first half of this long history, but three different systems were used during the second half, between 2000 and 1000 B.C. During this long span of time, the language changed from the Minoan one, which has not yet been deciphered, to an early form of Greek.

The earliest written script of Crete is called Minoan Hieroglyphic (Fig. 5.1). It is not the same as Egyptian Hieroglyphic texts, but it has a few similarities because both systems use images to write the language. From the number of signs that were used, we can tell that Minoan Hieroglyphic was probably a syllabary with a written symbol for each syllable in the Minoan language. Too few examples survive for it to be deciphered. In addition to the small amount of evidence with which to work, all of the texts are very short, and many of them are engraved on sealstones (so grammar is probably absent). The script was first used about the beginning of the Middle Bronze Age around 2000 B.C.

The second writing system used during the Minoan period is called Linear A. The older Hieroglyphic script was awkward to use because it was composed of complex images that were time consuming to draw. Linear A simplified these images and it added a few new ones to improve the way the script translated the language into written signs. It seems to have been invented in southern Crete, but, by the beginning of the Late Bronze Age, it was being used across the island. Although it has not been deciphered, more is known about it than is the case with the Minoan Hieroglyphic script, and because the numbers can be read, it provides information about several subjects, including the production and drinking of wine.

A clay tablet found at the palace of Zakros in eastern Crete shows a good example of Linear A (Figs. 1.2:36; 5.2). Like the Minoan Hieroglyphic writing, Linear A is a syllabary with signs for individual sounds (both vowels by themselves and a consonant plus a vowel). In most cases a sign would be used for each syllable in a word, and common products like wine and wheat were given a single sign (called an ideogram). Some of the signs are simplified versions of the ones used in Hieroglyphic, and others are new.

The tablet from Zakros illustrated in Figure 5.2 is actually the back side of an accounting document found in the palace. Its date is Late Minoan I, just after the middle of the second millennium B.C. It is one of several such tablets that were excavated several decades ago when the palace was uncovered. The town of Zakros, at the eastern end of the island of Crete, was both an administrative center and an important Minoan seaport for trade with the Near East and Egypt. Among the discoveries at this site were many items that came to Crete by sea from the eastern Mediterranean region, including copper ingots and elephant tusks.

Linear A tablets are rectangular slabs of clay that were carefully smoothed on both sides so that the writing could be scratched into the surface with a pointed tool called a stylus. The clay documents were simply the quick counts made on the spot by the accountants, who would later transfer the information onto permanent leather documents, which were stored in the official palace archives. Unfortunately, the permanent records were written on perishable materials, so they do not survive. Some clay tablets survive by accident because they were hardened into ceramic in the conflagration when the palace was destroyed by fire.

The tablet from Zakros records a large amount of wine of more than one type (Fig. 5.2). The inventory on the back is in three lines. The top line says something about the wine with two signs, and they are followed by the ideogram (AB 131) for wine and the number recorded, which is three (shown with three vertical lines). A horizontal line then ends this transaction, and the next entry takes up two lines. Again, two signs note

Figure 5.1. Sealstones of serpentinite from Knossos with Minoan Hieroglyphic signs. After Evans 1921–1935, I, 196, fig. 43.

something about the product being counted, and they are also followed by the wine sign and its number. In the Minoan number system, horizontal lines mean 10 and vertical ones mean one, so the total is 78 units. In comparison with the first class of wine, this is a very large number. The third class of wine, recorded on the third line, totals 17 units. The tablet does not state whether the count is by volume or by number of jars or whether this is an inventory or an incoming or outgoing shipment. The words that identify the different categories of wine are in the undeciphered Minoan language, but the tablet still provides important information. It shows that wines of several types were being made for the palace, and that the scribes had to keep track of how much of each type was available. This is certainly a lot of wine. It must have been a major commodity at Zakros. The wine records on the back side of this tablet complete the accounting of several other agricultural products that were being counted and recorded along with the wine.

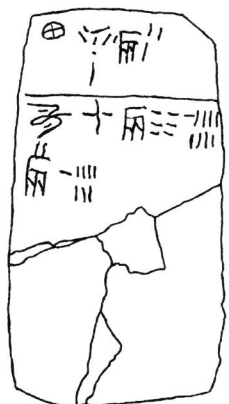

Figure 5.2. Clay tablet written in Linear A from the Minoan palace at Kato Zakros accounting quantities of different classes of wine. Drawing P. Betancourt.

The signs used for Minoan writing came from various sources. The ideogram for wine (AB 131) is unusual because it is one of the few elements with the same symbol used in all three of the Minoan writing systems (Fig. 5.3). The original sign was borrowed from the older Egyptian writing script (Egyptian Hieroglyphics). The Cretan sign is a slightly simplified version of the Egyptian one, depicting a series of grape vines. The original Egyptian image shows grape vines held up by forked stakes. The Minoans of Crete and the Egyptians of the Nile Valley were already trading at this time, and the information about a sign for wine must have reached Crete along this early trade route. The Minoans turned the sign sideways when they adopted it for their own writing system.

During the Late Bronze Age, the writing system of Crete was changed for a new script that was used to write an early form of Greek, a new language that was introduced to Crete after the Late Minoan I period. The new system is called Linear B. Because this script has been deciphered, we can read its messages. Linear B was used both in Crete and on the Greek mainland. It shows that by the Late Bronze Age, a Greek-speaking population, called the Mycenaeans after the city of Mycenae, had spread across the entire Aegean.

One of the new pieces of information provided by the Linear B documents involves the great expansion in wine production and consumption all over the Aegean region. The stability of the Mycenaean period created a large growth in population, and cities like Mycenae, Pylos, Sparta, and Thebes had kings who ruled over large regions. Great feasts and royal celebrations required thousands of wine goblets, and production of fine wines had to increase to support the new demands.

The Linear B tablets from Knossos do not list wine as something produced at the palace itself, but as an item that was received from elsewhere. The implication is that wine making in Crete during the Bronze Age took place near where grapes were grown, in many different places. It may have been a crop grown by independent landowners who tended

Figure 5.3. Comparison of ideograms used for wine in Egyptian Hieroglyphics and the three scripts used in Bronze Age Crete. Drawing P. Betancourt.

their own vineyards. This conclusion is supported by the many finds of pressing installations at different archaeological sites in Crete.

The Linear B tablets use many of the same signs as in Linear A, but using the script for a new language meant that many of the sounds were different, so new signs were needed as well. Like Linear A, the new script has very simple images that could be written easily and quickly. Most of the tablets were elongated instead of rectangular, as was usual for the earlier writing system.

One of the Linear B tablets from Knossos records that a large amount of wine was being shipped to Crete from the Cyclades. The reason for the large shipment is not stated on the clay tablet. The document is an accounting tabulation listing a total of 117 units of wine. It would presumably have been either a dedication to a deity at a religious sanctuary, or it would help supply a large celebration of royal proportions. Wine was now big business.

Further Reading

Evans, A.J. 1921–1935. *The Palace of Minos at Knossos* I–IV, Macmillan and Co., Limited.

Palmer, R. 1994. *Wine in the Mycenaean Palace Economy* (*Aegaeum* 10), Université de Liège and University of Texas at Austin.

Platon, N. 1971. *Zakros: The Discovery of a Lost Palace of Ancient Crete*, Scribner.

6

Classical Greece
Dionysos/Bacchus, the Greco-Roman God of Wine

Philip P. Betancourt

In the Greco-Roman world, the gods and goddesses were thought to have power over nature, controlling things like the weather, the growth of crops, and even the fates of human beings. One of the Greek deities was Dionysos (called Bacchus in Latin), a god who could affect agricultural crops and make the wine taste better. He was very cheerful, and he was particularly interested in celebrations with lots of wine and dancing. In ancient Greek art, he was always shown holding a large drink of a favorite vintage (Fig. 6.1), and he is often called the god of wine. He gradually became more influential in society, until the Roman period when his followers even believed the god could give them a life after death.

The main festivals in honor of Dionysos were held in the spring. Because they took place when the seeds were being planted for the year's new crops, honoring Dionysos was a request for a good harvest. The festivals were filled with hope, and they also coincided with the maturing of the wine from the previous fall, so that worship of the god became synonymous with sampling the quality of the new wine. With this cheerful god (Fig. 6.2), festivals were joyful affairs with free wine for everyone and much singing and dancing.

In Greek art, the god of wine can be recognized by his oversized drinking cup (Fig. 6.3). He had several followers. The female ones are called maenads, and the male ones, who are part animal and part human, are called satyrs, if they have goat tails and ears, and silens if they have horse tails and ears (Fig. 6.4). On Greek vases, these creatures seem to spend most of their time dancing and drinking wine. The followers of the god

Figure 6.1. Dionysos, the god of wine, holding a drinking vessel made from a cow's horn, painted on an Athenian black-figure vase attributed to Lydos. Ca. 550 B.C., height 56.4 cm, diam. 58.6 cm. New York, The Metropolitan Museum of Art, Fletcher Fund, 1931, acc. 31.11.11, https://www.metmuseum.org/art/collection/search/253349, accessed May 19, 2025; CC0 1.0, https://creativecommons.org/publicdomain/zero/1.0/.

Figure 6.2. A mask of Dionysos painted on an Athenian black-figure krater, a large vessel for mixing water with wine. Ca. 520–510 B.C., height 34.3 cm, diam. 33.3 cm. New York, The Metropolitan Museum of Art, Rogers Fund, 1906, acc. 06.1021.101, https://www.metmuseum.org/art/collection/search/247267, accessed August 1, 2025; CC0 1.0, https://creativecommons.org/publicdomain/zero/1.0/.

are often featured in humorous vignettes in Greek mythology (Fig. 6.5), and they always seem to be having a good time.

The story of the birth of Dionysos is an interesting myth that can give modern readers some insights into aspects of Greek thinking that are not always obvious. The god was the son of Zeus, the king of the gods, and a mortal woman named Semele. The mythological story says that Zeus would visit his lover in disguise as an older man. When his wife Hera became aware of the love affair, she went to visit Semele in disguise and tricked her into asking her lover to reveal his true identity by getting him to vow that he would grant her one wish. Even when he learned that it was a trick to reveal his true form, Zeus could not go back on his sacred vow to grant Semele her wish. Because Zeus is actually a lightning flash, not a human of flesh and blood, Semele was burned up when he changed into his real form, but her unborn child Dionysos was saved and sewn into the thigh of Zeus until he was born. Hera got her revenge on Semele who was killed and on her husband who lost his lover.

The story provides the modern reader with two insights into Greek mythology and art. First, it shows that, in spite of their art where the Greek gods look just like humans, sophisticated Greeks had a very different concept of their deities as metaphysical beings. And second, it points out that some of the myths were more like literary tales with morals than actual beliefs. Of course, the morals were quite chauvinistic compared to present day. In this case, the message was to blame the victim in the story, an attractive young woman who should not have granted sexual favors to an older man who unbeknownst to her had a possessive wife. A young woman is liable to get into trouble through no fault of her own. In fact, from our modern perspective, Zeus is the one at fault in the story by treating both women very poorly.

Dionysos became the god of the theater because his festival in Athens led to the foundation of theatrical performances. It was customary during the sixth century B.C. (called the Archaic period) in Athens to have professional singing and dancing groups perform at the festival of Dionysos to entertain those who attended the celebration. Three different groups were invited each year, and the audience voted to decide which one was best. In 534 B.C., the audience was amazed when Thespis, the leader of one of the performing groups, stepped aside from the rest of his troupe of singers and dancers and engaged in dialogue with them, using the performance to tell a story as if he were a character in the plot. The audience voted this group as the best, and actors are still called thespians to this day. The theater at Athens with its elaborate stone seats for important guests is an important monument (Figs. 6.6, 6.7). Its architectural form became the model for theaters elsewhere.

Many of our words about the theater are based on the original theater of Dionysos. Most of the surviving ancient theaters have been modified down the years, but the example at Epidauros (Fig. 6.8) preserves the

Figure 6.3. Dionysos on a red-figure pelike (jar) by the Geras Painter. The god holds an oversized drinking cup called a kantharos. Ca. 480 B.C., height 28.5 cm. New York, The Metropolitan Museum of Art, purchase 1901, acc. 01.8.8, https://www.metmuseum.org/art/collection/search/246933, accessed May 19, 2025; CC0 1.0, https://creativecommons.org/publicdomain/zero/1.0/.

Figure 6.4. Black-figure amphora featuring the god Dionysos holding his kantharos with two of his followers, a maenad at left and a silen at right. Ca 510 B.C., height 38.9 cm. New York, The Metropolitan Museum of Art, Rogers Fund, 1906, acc. 06.1021.85, https://www.metmuseum.org/art/collection/search/247252, accessed May 19, 2025; CC0 1.0, https://creativecommons.org/publicdomain/zero/1.0/.

Figure 6.5. Ceramic vase of a lounging figure holding a bunch of grapes and sitting beside an amphora, recognizable as a silen by his horse's ears. Fourth century B.C., height 11.3 cm. New York, The Metropolitan Museum of Art, funds from various donors, 1926, acc. 26.255.1, https://www.metmuseum.org/art/collection/search/252885, accessed May 19, 2025; CC0 1.0, https://creativecommons.org/publicdomain/ zero/1.0/.

Figure 6.6. Marble seats in the Roman Theater of Dionysos, Athens. Photo P. Betancourt.

Figure 6.7. Roman Theater of Dionysos, Athens. Photo P. Betancourt.

original design of its architecture. Performances are still held there every summer. The main parts of the theater are the auditorium, where the audience sat, the scene building that was the background (named after *skene*, the Greek word for a tent that was the original painted scenery for the performance), and the circular orchestra, where the dancing and acting took place. When a raised stage was added later, either musicians were placed in the orchestra (and given that name), or extra seats were added there as high-priced locations just in front of the stage.

Dionysos had festivals all over Greece, and the idea of theatrical performances quickly spread across the Hellenic world. Performances became a part of Greek society, and they were popular in many regions, including Crete. A good example is the ancient theater at Aptera, in the region of Chania in western Crete (Figs. 1.2:4, 6.9). The theater, which has been restored, lasted up until the Roman period, and it brought the plays of Aristophanes, Sophocles, Euripides, and other playwrights from Athens to the south Aegean island. The restored theater at Aptera shows the open-air architecture as it appeared in the Roman period at the end

Figure 6.8. Theater at Epidauros, Greece. Fourth century B.C. Photo P. Betancourt.

Figure 6.9. Theater at Aptera, Crete, with parts of the theater labeled. Fourth century B.C. Photo P. Betancourt.

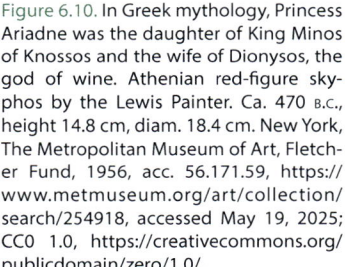

Figure 6.10. In Greek mythology, Princess Ariadne was the daughter of King Minos of Knossos and the wife of Dionysos, the god of wine. Athenian red-figure skyphos by the Lewis Painter. Ca. 470 B.C., height 14.8 cm, diam. 18.4 cm. New York, The Metropolitan Museum of Art, Fletcher Fund, 1956, acc. 56.171.59, https://www.metmuseum.org/art/collection/search/254918, accessed May 19, 2025; CC0 1.0, https://creativecommons.org/publicdomain/zero/1.0/.

of its history. The stone seats for the audience form a semicircle around the circular orchestra, where the earliest performances were located. The raised stage built across the edge of the orchestra was a later addition, allowing the actors to be better seen. As elsewhere, the performances always began with an offering to Dionysos as the patron of his festival.

In regard to Crete, the myth of Dionysos crossed over with the myth of Theseus, the great hero from Athens who killed the Minotaur. The mythological story says that King Minos of Knossos forced Athens to send seven young men and seven young women to Crete periodically to be fed to the Minotaur, a monster who was kept in a labyrinth. Theseus, the son of the king of Athens, volunteered to be one of those sent to be sacrificed. He killed the Minotaur with the help of the daughter of King Minos, named Ariadne (Fig. 6.10). The princess gave him a ball of twine to unroll as he went into the labyrinth, so he could follow the string back out and escape. Theseus and the princess escaped to a ship with the other young men and women from Athens and started to sail home. This is where Dionysos enters the story.

On the way back toward Athens, Theseus (who proved himself to actually be something of a cad and a scoundrel) fell in love with Ariadne's younger sister, abandoning the princess who had helped him and leaving her on the island of Naxos (Fig. 1.6) while he went back to Athens with the younger sister. Poor Ariadne was left alone, but the god Dionysos saw her and was instantly smitten with love. While the sister became the queen of Athens, Ariadne had a much finer future. She became an immortal goddess, and she and Dionysos lived happily ever after with lots of joy and much singing and dancing. So the myth had a happy ending for everyone.

7

Cretan Heritage 1
The Wine Glass with a Stem and Base

Philip P. Betancourt

One of the interesting details about drinking wine in the 21st century is that in most places in the Western World—from London to New York to Vienna to Athens—when you order your glass of wine, it will come in a Minoan vessel shape. One of the social manners we inherited from early times is that the container must fit the beverage. Does anyone drink coffee out of a crystal goblet? Certainly not in polite company! Coffee requires a coffee cup, as hot tea requires a teacup, and iced tea requires a glass. Wine is usually served in a goblet supported by a stem rising from a circular base. This container is a Minoan shape, inherited from examples used well before 2000 B.C. on the island of Crete. It is one of the Bronze Age ideas that is still with us, like building a town with an open communal square around which are located the important buildings where the citizens need to gather.

The shape was already present in some parts of Crete during Early Minoan I (just after 3000 B.C.), but it had to compete with other vase shapes before it emerged as the surviving social custom for drinking wine. In several Minoan towns at the beginning of the Bronze Age, when someone died, he or she would be commemorated with a final toast at the cemetery. The toast was made with a gigantic wine goblet that was shared by the entire group of mourners (Fig. 7.1). The vessel must have been passed from person to person among the guests. At the cemetery of Hagia Photia, we can see that the goblet was then smashed and left on the grave as a final tribute to the deceased. One suspects that the large goblets were not filled all the way to their brims because a large empty space above the

Figure 7.1. Large communal goblet made of Pyrgos Ware from the Hagia Photia cemetery, Crete. Early Minoan IB, ca. 2800 B.C., height 21.6 cm. Photo P. Betancourt.

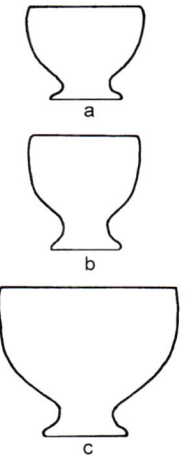

Figure 7.2. Three goblets of Vasiliki Ware from Vasiliki, Crete. Early Minoan II, ca. 2600 B.C., scale 1:3. Reprinted from Betancourt 2007, 42, fig. 3.15.

Figure 7.3. Goblet from Mycenae, Greece, ca. 1400 B.C., height 16.3 cm. Reprinted from Betancourt 2007, 157, fig. 8.1.

liquid would have held the wine's fruity aroma and added to the pleasure of drinking. As the groups of mourners grew larger, the ceremony took too long, so in Early Minoan II it was changed in order for everyone to get their own cup, which could be filled from a jug in preparation for the toast. The individual smaller cups were modeled after the larger communal ones, and they were goblets made with a stem and base (Fig. 7.2). The new custom improved the ceremony because now everyone could drink at the same time, just after the leader announced the toast.

The use of a goblet for wine drinking was not yet universal throughout Crete. It was the fashion at Knossos and several other places, but the people of Phaistos still preferred simple conical cups without bases as late as the Middle Bronze Age, and many people in the Pediada region of central Crete preferred tall tumblers without either handles or bases.

By the Late Bronze Age (ca. 1500 B.C.), the goblet was part of an international style of fine pottery in the Aegean. Knossos had extended its political control (and many of its customs) over most of the island of Crete. Soon the Mycenaeans from mainland Greece would take up the Cretan customs when they established their dominance across both Crete and the rest of the Aegean, with their trade extending to both the east and the central Mediterranean. Aegean pottery became international. For fine drinking, the goblet now had a pair of handles added to the tall shape with its stem and base so that the host could hand the vessel to a guest (Fig. 7.3). In some places, the stem gradually became much taller (as in Fig. 7.4), giving the vessel an even more stately appearance. The elegant variations with stems, bases, and handles were popular for many years, until a serious new challenge to the drinking shape appeared around 1200 B.C. at the end of the Bronze Age.

The final years of the Bronze Age were a troubled time throughout the whole Aegean region. The palaces of both Crete and mainland Greece were destroyed, social order broke down, the seas became so unsafe that most coastal settlements moved inland, and many people left the entire Aegean for Cyprus and other places. It is interesting that even a social custom like what to choose as a drinking cup was affected. The goblet received some serious competition! A drinking bowl with a wide rim and two horizontal handles, called a skyphos appeared (Fig. 7.5).

The new drinking vessel was not popular everywhere, and some regions continued to use tall drinking cups with vertical handles. The potters solved the problem by combining the main aspects of each cup to create a new class of kylix (Fig. 7.6), which had the shallow bowl of the skyphos with its horizontal handles as well as the stem and base of the goblet. The new invention was popular, and it became a favorite vessel for drinking wine in Athens throughout the Archaic and Classical periods.

Eventually, the Greek tradition of the kylix with a shallow bowl declined in popularity, and when blown glass drinking vessels came along at the beginning of Roman Imperial times, the tall shape completely replaced the shallow one, and handles tended to disappear as well. The

Figure 7.4. Mycenaean kylix with stylized flower. Late Helladic IIIB:1, ca. 1300–1225 B.C., height 19.6 cm, dia. 17.1 cm. New York, The Metropolitan Museum of Art, gift of the Greek Government, 1927, acc. 27.120.8, https://www.metmuseum.org/art/collection/search/252906, accessed August 1, 2025; CC0 1.0, https://creativecommons.org/publicdomain/zero/1.0/.

Figure 7.5. Athenian skyphos decorated with lines, chevrons, and dots. Ca. 800–750 B.C., height 6.8 cm. New York, The Metropolitan Museum of Art, gift of the American Society for the Excavation of Sardis, 1926, acc. 26.199.286, https://www.metmuseum.org/art/collection/search/252849, accessed May 19, 2025; CC0 1.0, https://creativecommons.org/publicdomain/zero/1.0/.

Figure 7.6. Athenian black-figure kylix with siren and "panther." Attributed to Tleson as painter and potter. Ca. 550–540 B.C., height 10.3 cm, diam. 14.1 cm. New York, The Metropolitan Museum of Art, gift of F.W. Rhinelander, 1898, acc. 98.8.16, https://www.metmuseum.org/art/collection/search/246728, accessed May 19, 2025; CC0 1.0, https://creativecommons.org/publicdomain/zero/1.0/.

Gothic chalice (Fig. 7.7) continued the tradition of a tall bowl supported by a stem rising from a circular base. It belongs to a class of chalices used in church services in commemoration of the last supper of Christ with the chalice as the container for the holy wine.

Goblets with stems and bases have continued to be used up until the present day. As an example, a goblet manufactured in Venice around 1960 (Fig. 7.8) is made of very fancy glass with gilding and flowers in colored glazes. It has an elaborate decoration but a traditional overall form. Modern wine glasses come in various shapes, but the stem and base to support the container are survivals from their Minoan ancestors.

Good reasons explain why the details of a shape with a stem and base have been successful parts of a wine glass for a long time. They allow the vessel to be easily twirled to swish the wine and produce its pleasant aroma, and the rounded bowl holds the scent longer if it has empty space above the liquid. If held by the stem, the heat of one's hand does not change the temperature chosen for the vintage, and a desirable elegance is projected by the tall and attractive form. The details of stem and base are well suited to the drinking of this pleasant beverage, all thanks to the ancient Minoan shape.

Further Reading

Betancourt, P.P. 2007. *Introduction to Aegean Art,* INSTAP Academic Press.

Figure 7.7. Gothic chalice with a rounded bowl supported by a stem and base; possibly from Hungary. Silver decorated with gold filigree, enamel, glass, and possibly semi-precious stones. Mid 15th century, height 21.6 cm, diam. 11.2 cm. New York, The Metropolitan Museum of Art, gift of the Salgo Trust for Education, New York, in memory of Nicolas M. Salgo, 2010, acc. 2010.109.7, https://www.metmuseum.org/art/collection/search/478974, accessed May 19, 2025; CC0 1.0, https://creativecommons.org/publicdomain/zero/1.0/.

Figure 7.8. A violet-tinted glass goblet with additions of gold and small flowers made in a workshop on the island of Murano, Venice, Italy. The pedestal rises from a disk base and supports the rounded bowl. Gilded on the upper part of the exterior along with raised gold vine-like tendrils and flowers made of pinkish-white, pink, and blue glass along with green and white glass leaves. Height 13 cm. Acquired in 1961. Photo P. Betancourt.

8

Cretan Heritage 2
The Great Jar Tradition

Philip P. Betancourt

Agricultural communities have always developed traditions that suited the materials they had available and the opportunities and challenges provided by their natural resources. The communities of Crete are a fine example. To mature properly, wine needs to be stored for a length of time in a water-tight container that will prevent spoilage from moisture or microbes or small animal life like insects or rodents. For the early Cretan settlers who lived in houses with soil floors and walls constructed of stones held together by mud mortar, secure protection from insects and rodents over an extended period of time needed a material that was more durable than basketry, leather, or wood. Beginning already at the end of the Neolithic period before 3000 B.C., the residents of Crete began making large clay vessels for the storage of wine and other commodities from one harvest to the next. Any Cretan jar made in the Bronze Age that was over 50 cm high can be called a pithos. The jars were used to store many different items, and they were especially useful for wine.

Creating large clay vessels that would hold significant quantities in a water-tight and vermin-proof container was a technological challenge for early potters. The manufacture of a large jar was simple enough, but the firing of an especially large object in a potter's kiln required both specialized knowledge and experience. The problem is that the temperature must be wholly uniform around the item during the entire period of the firing, because the rising temperature will cause successive changes in the object as gases are lost and the clay materials experience changes in both size and composition. Stresses from different temperatures on a single object will cause it to break. In the early kilns of Crete, the rising temperature was achieved slowly and carefully, with kilns that had rounded interior corners to avoid the problem of trapping the air in tight spaces. Large jars were made of clay recipes that were coarse enough to permit the gases to escape easily during firing and thick enough to make a very durable end product. This made the jars rather heavy, even when they were empty, and they were seldom moved after they were full.

The jars have two names. The Bronze Age version is called a pithos (pithoi in the plural), and the Middle Ages to modern version is called a pithari (pitharia in the plural). They have been made in Crete for over 5,000 years, not necessarily in every generation (given the durability of the pithoi), but, when needed, a skilled potter could always look at an old example and figure out how to make another one. They are extremely stable, and, except for an earthquake or a house fire, not much harm can come to an example sitting quietly in its storage area.

The earliest Cretan pithoi (Figs. 1.4, 1.5) come from the end of the Neolithic and the beginning of the Bronze Age. Their inspiration came from the Cycladic islands where they had already been manufactured for some time. Early potters in the Siteia region of northeastern Crete became very skilled at firing these jars, and most of the earliest versions were made there and transported elsewhere by sea. By the beginning of the Bronze Age, the recently domesticated donkey enabled overland travel and the movement of heavy goods, a great expansion in trade.

Among the many different jar designs that were developed in Crete, coils of clay added to the exterior surface became the most popular decoration for the large containers (Fig. 8.1). The added clay bands were use-

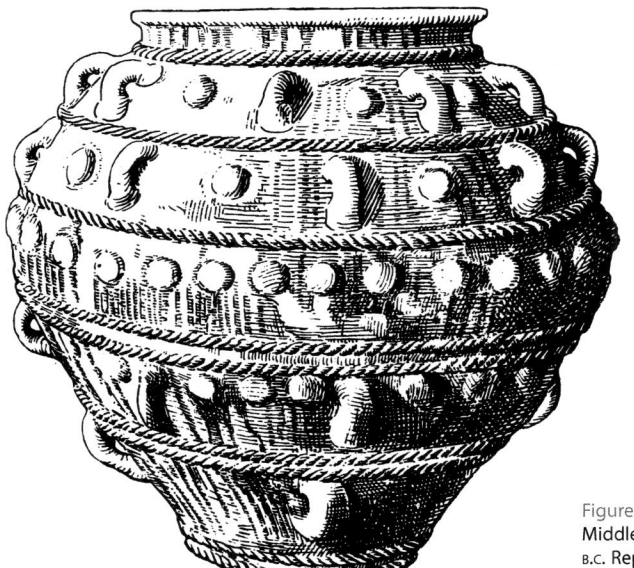

Figure 8.1. Pithos from Phaistos. Middle Minoan II, ca. 1875–1750 B.C. Reprinted from Evans 1921–1935, I, 232, fig. 174.

Figure 8.2. Excavations of the House of the Metal Merchant at Mochlos uncovered several pithoi. Late Minoan IB, ca. 1625–1470 B.C. Photo courtesy J. Soles.

ful in two ways. First, they helped strengthen the walls during the drying and firing process by giving support without adding to the thickness and the weight of the entire vase. And, secondly, they allowed the finished jars to be recognized individually without opening them (at least as long as the owner could remember what had been stored in each one). The fact that not every house in a community had them suggests that cooperation and community storage existed in villages where several of the households would be related. The Cretan rulers built large architectural complexes as their official palaces, and these buildings included extensive storage magazines where the wine and other agricultural crops were preserved as part of their official wealth. As an example, the storage magazines at Knossos on the western side of the central court with their rows of pithoi were large enough to hold supplies for many months (Fig. 1.5).

Numerous other examples exist. A group of large pithoi were found in a Late Bronze Age store room at the settlement of Mochlos in northeastern Crete (Figs. 1.2:22, 8.2). This find is one of several discovered in the excavations at Mochlos. They come from the House of the Metal Merchant. The owner of the house was a wealthy resident of the port town. The building included many copper and bronze objects, including ingots that had been imported from Cyprus as well as the pithoi to store the food supplies needed for the household.

In the middle centuries of the second millennium B.C., Minoan potters blurred the lines between their large storage jars and their fancy dining ceramics by decorating large vessels for display. Pithoid jars, like an example from Pseira, were covered with fine clay slip so that their decoration could be of the highest quality (Fig. 8.3). The vase is painted with a field of spirals. Another fine jar has a lively octopus whose wavy tentacles mimic the display of marine life (Fig. 8.4). Such jars must have been placed in the palace in locations where their large size and attractive decoration could be seen by visitors.

The making of large jars continued in Crete for many centuries after the Bronze Age, providing a service to a population that was mostly agricultural. The tradition from Medieval to modern times can be traced

Figure 8.3. Jar from Pseira decorated with a field of spirals. Late Minoan IB, ca. 1625–1470 B.C. Photo P. Betancourt.

Figure 8.4. Jar from Knossos decorated with an octopus. Late Minoan II–III, ca. 1470–1075 B.C. Photo P. Betancourt.

through the work of the potters from Thrapsano, known until recently as the "village of jar-makers."

Thrapsano is a town in central Crete (Fig. 1.2:31). Until the mid 20th century, it was the home base for 30–35 pottery workshops making the large jars called pitharia. The workers were farmers for part of the year, but each spring teams of six men would leave the village and travel with their tools to locations around Crete where they would set up their workshops to make jars for the people of the local region. The workshop team members included:

- The master, who was in charge of the workshop.
- The second master, his chief assistant, who also helped manage the kiln.
- The wheeler, who rotated the turntable for making the jars.
- The clay man, who was in charge of digging and preparing the clay.
- The wood cutter, who cut the wood for the kiln.
- The carrier, who transported the clay, fuel, and finished jars using donkeys.

Two clays were mixed to make the correct fabric recipe for the jars. One of them was a red clay that was firm but too difficult to work by itself, and the other one was a pale brown clay that was too pliable to use by itself. For success, they had to be mixed. The Cretan clays are a little different from place to place, and the proper mix of the two varieties need to be adjusted to suit the local conditions. All the materials needed were available for free, and the workshops were set up in the countryside where land was available without cost.

The workshop was mostly outdoors. The jars were made on turntables placed in a long trench so the base of the jar would be low and at ground level, and the rim would not be too high to reach easily while sitting down. A series of 10 to 16 side compartments is cut into the soil, with a turntable in each compartment. The turntables were circular disks made mostly of wood. These flat pottery wheels were attached to central axles, and they rotated easily as each one was firmly tied to horizontal crossbars with oily strips of cloth (Fig. 8.5). Each turntable was used to make its own jar. One set of pitharia was made each day, with the master and the wheeler moving down the line of turntables and working on each one in succession (Fig. 8.6). The potters added one ring of clay at a time and allowed the clay to dry in between each stage while the pair of workers moved down the line to the next turntable. After building up the jars with six successive rings of clay, the jars were finished with a thick rim (that proved useful later in moving the jar), a set of handles that helped tie it in place on a donkey, and some decorative bands.

At the end of the day, the potters would have finished the whole set of jars in the trench, and they would be left in place to dry overnight before removing them from their turntables the next morning. The jars needed to dry several additional days before they were ready to be fired in the kiln. The other specialists in the workshop were busy with other tasks while the two masters and the wheeler made the jars. The kiln and its stock of wood for fuel would be ready when enough jars were dry to fill it.

The jars were fired in a two-story kiln with the room full of jars above a space for the fire. A clay floor with holes in it to allow the heat to rise separated them. The favorite fuel was the seeds and skins left over from pressing olives to make oil.

The pitharia would all be sold or exchanged for food and other supplies. When some jars were sold to a customer, they were loaded on a donkey, two jars for each trip, and they were delivered to the purchaser's home. If anyone needed new jars (they were only sold in pairs), they

Figure 8.5. Traditional turntable used by potters from Thrapsano. Photo courtesy M. Voyatzoglou.

Figure 8.6. Potter from Thrapsano building a jar on a traditional turntable set in a trench. Photo courtesy M. Voyatzoglou.

Cretan Heritage 2: The Great Jar Tradition | 47

Figure 8.7. Potter in 2022 using an electric motor to rotate the turntable. Photo P. Betancourt.

Figure 8.8. Modern jars produced in Thrapsano. Photo P. Betancourt.

would have to purchase them during the summer when the potters were present in the region.

By the middle of the 20th century, containers made of new materials like plastic were available for secure and safe storage, and supermarkets with year-round imports of foods meant that long-term storage was no longer essential. Houses no longer had dirt floors and walls of stones held in place with mud mortars, so the homeowner no longer depended on this type of storage container for safety. The need for the large pitharia as a form of agricultural storage units declined rapidly. Only a few workshops survived the radical changes in the new economy. They stopped traveling, set up permanent workshops in Thrapsano, and broadened their output to produce flower pots and other vessels as well as pitharia.

Many jars are still being made in the 21st century as decorative pieces, but the methods have changed. The line of turntables no longer needs to be in an excavated trench out of doors. The potter's wheel is no longer a handmade device with a vertical spindle tied to a wooden strut with an oily rag. Well-made electric machinery replaced the wheeler, who once needed to squat in the trench and move from turntable to turntable (Fig. 8.7). Even with all these modern developments, the clays are still the same, and the skilled master potter can still make whatever shape is desired. Modern sales are international, and the finished jars are still made in large numbers and shipped wherever the market has enough customers (Fig. 8.8). In addition, if one travels to the beautiful island of Crete, some of the older jars can occasionally still be seen at garden parties (Fig. 8.9); one change is that they no longer hold wine.

Further Reading

Christakis, K.S. 2005. *Cretan Bronze Age Pithoi: Traditions and Trends in the Production and Comsumption of Storage Containers in Bronze Age Crete* (*Prehistory Monographs* 18), INSTAP Academic Press.

Evans, A.J. 1921–1935. *The Palace of Minos at Knossos* I–IV, Macmillan and Co., Limited.

Voyatzoglou, M. 1984. "Thrapsano, Village of Jar Makers," in *East Cretan White-on-Dark Ware: Studies on a Handmade Pottery of the Early to Middle Periods* (*University Museum Monographs* 51), P.P. Betancourt, University Museum, University of Pennsylvania, pp. 130–142.

Figure 8.9. Garden party at the Tholos Beach Hotel, Kavousi, Crete, with Susan Ferrence and Alessandra Giumlia-Mair admiring a pithari (storage jar) from Thrapsano painted with decoration and providing an attractive accent in the garden. Photo P. Betancourt.

9

Byzantine Wine and the *Geoponika*

Philip P. Betancourt

The *Geoponika* is a large encyclopedia of Byzantine agricultural practices written in the 10th century during the reign of Emperor Constantine VII Porphyrogenitus. It consists of 20 books about farming and animal husbandry with sections on different aspects of food production. Its information draws on a long tradition of earlier writing about agriculture, especially the work of Cassianus Bassus, who had written a volume with this name in the sixth century. Two books discuss growing grapes, and three books are on making wine. Additional topics include growing grains, olive cultivation, animal husbandry, beekeeping, building ponds for fish, keeping pigeons and other birds, and many other subjects about farming and farm animals.

The work was commissioned by the Byzantine emperor as an attempt to gather knowledge about various topics from scattered sources in several languages. It was a worthy scholarly effort at a time when many of the sources were rare and difficult to obtain. The care lavished on vessels used for drinking wine during this period (Fig. 9.1) demonstrates the importance of the subject during the Byzantine era, and the topic of grape growing and turning crops into wine forms a major part of the treatise, with 25% of the books in the *Geoponika* devoted to this subject.

The *Geoponika* is important in the history of wine making for several reasons. It is very detailed. It draws on a great many previously written treatises on growing grapes and making wine. It is a substantial compilation of the practices before the revolution in wine making caused by the wider use of wooden barrels and the invention of mass-produced glass bottles altered the previous practices based on clay jars. In the 10th century in the Byzantine Empire, wine was still mainly stored in ceramic containers, and it required rather different production methods in comparison with more modern practices.

At the same time, the *Geoponika* has some problems. It was compiled from literature, not from first-hand observations, and it did not evaluate whether what it was reporting was actually true. Although it mentioned various sources that were used for the information it recorded, it did not state exactly what came from each source, so the reader is unable to understand the age or the origin of the practices mentioned.

As a result of the methodology, the information is very uneven. For example, in Book VIII, the reports about medicinal wines were never tested or evaluated by the writers of the *Geoponika* to see if they were correct, possible, or utter nonsense. In one passage the book says that putting an iron ring into the wine will keep it from turning sour, and, in another place, it suggests writing a religious inscription on the jar for the same purpose. Both of these methods are unlikely to do what they say and could have been tested easily, but they were only included because they were mentioned in the existing literature. The *Geoponika* was not a manual for agricultural practices, but a compilation of what had been written about them. With this purpose clearly in mind, it nevertheless offers a great amount of historical information.

Constantine VII Porphyrogenitus was the son of Byzantine Emperor Leo VI. As a young boy, the former was crowned co-emperor with

Figure 9.1. The Antioch Chalice, a plain silver goblet enclosed by an elaborate gilded shell with rinceau decoration: grape vines, birds, animals, and seated men. Ca. A.D. 500–550, height 19.6 cm, diam. 18 cm. New York, The Metropolitan Museum of Art, the Cloisters Collection, 1950, acc. 50.4, https://www.metmuseum.org/art/collection/search/468346, accessed May 19, 2025; CC0 1.0, https://creativecommons.org/publicdomain/zero/1.0/.

his father and his uncle Alexander in Constantinople in the year 911. He was only six years old and too young to rule when his father passed away the next year and his uncle became emperor. When Alexander died in 913, Constantine was again confirmed as the co-emperor in order to assure the succession. He was young and shy, and he did not really assume any duties until 931, after spending many years in obscurity, passing his time with literature and music rather than ruling an empire. During his reign, he assembled several teams of writers to record encyclopedic volumes about different subjects, and he is best remembered for these literary achievements. The *Geoponika* is one of these works.

Many topics about grape cultivation and wine are covered. Subjects include what land is appropriate, transplanting and cultivating the grapes, good and poor climate issues, time for planting, care of vines, staking the plants, and other needs. It includes a diary of what is to be done each month of the year. Making the wines is covered in detail, using clay jars of many sizes.

Much of the advice about making wine concerns cleanliness in order to eliminate problems and contamination. Grape leaves must never get into the presses. The vats are to be cleaned with sea water or prepared brine and dried for 20 days after cleaning before they can be used. Those who tread on the grapes have to keep their feet clean; they cannot eat or drink when pressing on the grapes, and they cannot get in and out of the vat used for the pressing and move around with bare feet.

Long sections of the *Geoponika* cover the fermenting and aging stages. Keeping the wine stable during this long period is a lengthy subject in the treatise. Ways of avoiding changes in temperature are the main concerns. Creative advice includes putting the wine jars in a well or a fountain, burying the vessels in sand, and other similar ways to keep the temperature stable.

Extensive directions are given to make the wine jars waterproof with resins. The resins are to be mixed with beeswax, wood ash, and other substances and then boiled before being added both to the interiors and exteriors of the storage jars. Both pine resin and terebinth are mentioned. Many flavors and medicines are discussed as additives to the beverage at this stage, and they can either be added to the wine directly or mixed with the resins so that they will be absorbed. The mouths of the jars are also sealed with the resin mixture.

The inclusion of ash in the waterproofing material is an important aspect. Wood ash, especially if made from plants that grow in alkali flats, contains sodium or potassium hydroxide, which are bases that neutralize acidity. The intent must have been to have an alkali present during the fermentation stage to reduce acidity and slow the conversion of wine to vinegar. It would improve the quality of the wine substantially.

The long treatise is uneven in content because it is a mixture of serious information and useless advice. It has detailed instructions on getting a test sample from the sediment at the bottom of a jar by using a hollow reed without mixing the sediment with the good wine, but it also advises putting flowers on one's head to avoid intoxication. The *Geoponika* has a wealth of information on grapes and wine, but it has to be read judiciously.

Further Reading

Dalby, A. 2011. *Geoponika: Farm Work. A Modern Translation of the Roman and Byzantine Farming Handbook*, Prospect Books.

10

The Enigmatic Malvasia di Candia Wine

Albert Leonard, Jr.

Malvasia di Candia, a very popular sweet wine, has been associated with the island of Crete/Candia since at least the 14th century A.D. During the many years since then, there has been considerable disagreement about the meaning of this wine's name. Does the name indicate the place where the wine was made, or does it identify the grape from which the wine was crafted? At the root of much of this confusion is the fact that the word Malvasia (or a derivative of it) appears in the names of more than 20 unrelated grape varietals. These grapes are grown all around the Mediterranean and even beyond (Robinson, Harding, and Vouillamoz 2012, 571–587). Let us first look at the geographic element and then examine what *ampelography* (the identification and classification of grapevines) can tell us.

Malvasia the Place

Two basic interpretations exist today for the word Malvasia: it is either a port in Laconia on the Greek mainland (in the Peloponnesus) or an agricultural region on the island of Crete. The questions arise about the relationship (or lack of a relationship) between these two locations.

According to one view, Malvasia represents a translation, contraction, variation, or perhaps some other relation to the modern Greek town named Monemvasia, a fortified port near the southern tip of the mainland (Figs. 1.6, 10.1). Literally, Monemvasia means one entrance (*mono-vasia*) with reference to the single bridge that connects the island to the Peloponnesian mainland, thus creating one harbor to the north and a second harbor to its south.

To a modern casual traveler, Monemvasia is a rather isolated place: a brief stop on a six-hour ferry ride from Athens's port of Piraeus before the ship travels on to the islands of Kythera, Antikythera, and, finally, the port of Kastelli Kissimos on the western coast of Crete (Fig. 1.6). But that was not the case in antiquity. Except for the canal at Corinth (opened in 1893) and an earlier paved trackway (the *diolkos*) across the isthmus (Fig. 10.2), it was nearly impossible to move people or products between the Aegean and the Ionian Sea without sailing around stormy Cape Malea (Μαλέα), the tip of the easternmost finger of mountains pointing out from the Peloponnesus southward toward Crete (Fig. 1.6). The dangers inherent in making a safe passage around a promontory whose Greek name suggested the shrillness of its winds were legendary. When Odysseus attempted it on his way home from the Trojan War, he was blown southward past the island of Kythera, all the way to the land of the mythical Lotus-eaters (Homer, *Odyssey* 9.90). Centuries later, during the reign of the emperor Augustus (r. 27 B.C.–A.D. 14), the strength of those winds prompted the Greek geographer Strabo (63 B.C.–A.D. 23) to caution grimly that "when you round Cape Malea, forget about your home" (Strabo 8.6.20).

Monemvasia is situated about 20 km north of that treacherous cape in an area that was known as *Epidaurus Limera* during Byzantine times. It forms one of the most distinctive topographical features in the Mediterranean. With a fortified summit towering more than 200 m above sea level, it has served for centuries as an important landmark to guide navigators to the double port below. Here the captain of a westbound ship could take on provisions and cargo, while its crew prepared the craft for

Figure 10.1. The island of Malvasia as it appeared in the 17th century when it was ruled by the Ottoman Empire. The bridge that connects the island with the mainland is visible at left. Map F. de Witt, Amsterdam, 1680, https://commons.wikimedia.org/wiki/File:De_wit_1680_monemvasia_b.jpg#/, public domain, accessed May 21, 2025.

Figure 10.2. The Corinth Canal follows much of the paved (cobblestone) track (*diolkos*) that was begun at the end of the 7th century B.C., connecting the Aegean and Ionian Seas. Inaugurated in 1893, the canal measures 6.3 km long and saves ships 185 nautical miles and the possibly stormy trip around Cape Malea. Photo courtesy Dingoes67, https://commons.wikimedia.org/wiki/File:Corinth_Canal_with_boat.jpg#/media/File:Corinth_Canal_with_boat.jpg, accessed May 21, 2025; CC BY-SA 4.0, https://creativecommons.org/licenses/by-sa/4.0/.

the stormy passage ahead. It also became the perfect place to celebrate a successful eastbound passage—perhaps with a cup or two of its famous Malvasia wine (Μονεμβάσιος οίνος, *Monemvasios oinos*).

The alternate view of the origin of the word Malvasia is that it refers to Malevizi, a grape-producing municipality in central Crete (at present in the Herakleion regional unit). Here in the 1900s three palatial Minoan villas were excavated by Greek archaeologists at the site of Tylissos (Fig. 1.2:30). Domaine Zacharioudakis Winery is in this area, and it presents a very detailed case in favor of this alternate view on its web page (zacharioudakis.com). It is supported by bibliography and references to several antique maps of Crete that are now in the Benaki Museum in Athens. Most of the historical sites relevant to this discussion have been chronicled (with photographs) by Stelios Manolioudis in his *Wine Routes in the Cultural Landscapes and Malvasia of Crete* (see below, Further Reading). The *Malevyziotikos*, a very popular and strenuous dance, is still at home in the area, as are several other excellent wineries, including Diamantakis Winery, the Idaia Winery, and Silva-Daskalakis Winery, to name a few.

There are a few other hypotheses on the origin of the name Malvasia, but these are the main two. And keep in mind that selecting one need not exclude the other.

Malvasia the Grape(s)

About 50 different grape varietals include the word Malvasia (or a variant) in their name. These vary according to the language employed: Malvasier, Malvoisie, Malvagia, and Malmsey are some of the more common. Often, they are further modified by a placename (di Candia, di Lipari, etc.) or a descriptive adjective (Bianca meaning white, Nera meaning dark, Aromatica meaning fragrant, etc.). These grapes display a wide geographic distribution across the Mediterranean world, even extending a considerable distance out into the Atlantic, where a grape known as Malvasia Malmsey creates the sweetest of the sweet wines produced on the island of Madeira (southwest of Portugal). DNA analyses have shown that these names are not synonyms for the same grape, but rather they represent separate, distinct varieties each with its own complex genealogy and, as such, they are often discussed collectively as "The Malvasia Group," and individually as "Malvasia Somethings" (see Robinson, Harding, and Vouillamoz 2012, 571).

At first glance, three of these Malvasia Somethings would appear to be associated with the island of Crete/Candia. As each is discussed below, keep in mind that Crete is the fifth largest island in the Mediterranean, and it covers a land mass of over 8,000 square km (over 3,000 square miles). Also consider the difficulty in setting specific chronological brackets around a place name such as Candia that traces its roots back to the 820s when marauding Saracens named their new capital Rabaḍ al-Khandaq (Fortress/Castle of the Moat). That name was assimilated into other languages during subsequent periods of occupation. To the Byzantines, it was Chandax or Chandakas. At the beginning of the 13th century, however, after the island had become a possession of Venice (Fig. 10.3), its name was Italianized to Candia (or Candie), referring first to the town and, eventually, to the whole island. As part of the Venetian Kingdom (or also called Duchy) of Candia, the name could even include neighboring Kythera and the far-away island of Tinos (Fig. 1.6).

Figure 10.3. The fearsome winged Lion of St. Mark, depicted on the wall of the Venetian fortress in the harbor of Herakleion and the symbol of the Republic of Venice, stands ever vigilant to protect the island of Crete. His Bible is held open to the greeting of the angel, *PAX TIBI MARCE EVANGELISTA MEVS* (May peace be unto you, Mark, my Evangelist), who promised that one day the saint would be buried in Venice. Photo courtesy A. Leonard.

With baggage such as this, a grape variety that includes Candia in its name might not be as specifically Cretan as it first appears. The three relevant Malvasia Somethings are Malvasia Bianca di Candia, Malvasia di Candia Aromatica, and Malvasia Cândida.

Malvasia Bianca di Candia (Fig. 10.4) can cause considerable confusion on wine labels when it appears simply as Malvasia di Candia or Malvasia Candia without including the word bianca. The grape is thought to be a descendant of an old Italian varietal known as Garganega that either originated in Greece or somewhere farther east in the Levant. Genetically, Malvasia Bianca di Candia is identical to the Grecanico Dorato grape that is grown extensively in Sicily, both in the eastern part of the island, on the slopes of Mt. Etna, which was settled by Greeks in the 8th century B.C., and, especially, in the western part, which was colonized by the Phoenicians about the same time. The fact that the Malvasia Bianca di Candia grape naturally produces a sweet wine might account for its association with the island of Crete where such wines were historically favored.

A monovarietal wine of Malvasia Bianca di Candia is released by Idaia Winery: *Malva de Crete* (12.5% abv.). Miliarakis Winery (Minos brand) offers a blend of 50% Malvasia Bianca di Candia mixed with 50% Malvazia Aromatica in its dry *Malvasia* white wine (12.5% abv.).

Malvasia di Candia Aromatica is a separate and distinct grape variety and not simply a more aromatic variant of the Malvasia Bianca di Candia grape. Often referred to simply as Malvasia Aromatica, it is encountered today in both Greece and Italy. DNA analysis suggests that the grape is

Figure 10.4. The Malvasia Bianca di Candia grape. Photo courtesy U. Brühl, Julius Kühn-Institut, Federal Research Centre for Cultivated Plants, Institute for Grapevine Breeding Geilweilerhof, Siebeldingen, Germany, https://commons.wikimedia.org/wiki/File:VIVC23555_MALVASIA_BIANCA_DI_CANDIA_Cluster_in_the_laboratory_9629.jpg, accessed August 1, 2025; CC BY-SA 4.0,https://creativecommons.org/licenses/by-sa/4.0/.

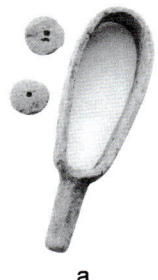

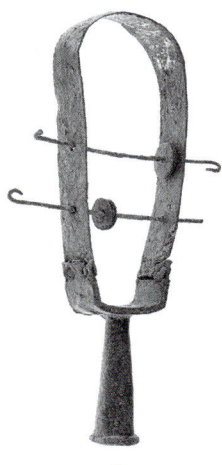

related to the Malvasia di Casorzo grape grown in northern Italy. Malvasia Aromatica is cultivated at Scalarea Estate (at Fantaxometocho near Herakleion). On Crete the grape is used to produce both monovarietal wines and several interesting blends.

Monovarietal wines made of Malvasia di Candia Aromatica are more common than blends on Crete. Douloufakis Winery releases two examples: *Femina* (13.9% abv. with 2.4 g/L of R.S.) and a *vin de liqueur* (16% abv.) presented in a 500 ml bottle. Dourakis Winery offers *Zelos Aromatica Bio* (13% abv.). Titakis Winery uses 100% sun-dried grapes in its naturally sweet *Arcano* wine that spends 30 months in oak (at 13.4% abv. with 110 g/L of R.S.), also in a 500 ml bottle. Domaine Gavalas offers *Gavala's Malvazia*, and Pnevmatikakis Winery produces a semi-sweet *Afronas Chronos Reserve* white wine (12.5% abv.).

Malvasia di Candia Aromatica is also used in blends. Diamantakis Winery adds it to a half and half blend of Chardonnay in its *Prinos* white (12.8% abv.). Efrosini Winery uses 40% Assyrtiko to make its *Onirikon* (Dreamlike) white wine (13% abv.). Titakis blends in Vidiano to achieve its *Impetus White*, a dry wine that spends four months in oak (12.5% abv.). Agelakis Winery uses a proprietary percentage of Vilana in its dry white *Seistro* (*Sistrum*) blend that depicts a sistrum on its label as a memento of Minoan heritage (Figs. 10.5, 10.6).

Malvasia Cândida is not a spelling variant of either of the two grapes discussed above. It is a separate varietal that is related to the Malvasia di Lipari grape whose origins are also cloudy. Archaeology has shown that the island of Lipari (specifically; Fig. 1.7) and the volcanic Aeolian Islands (in general) were visited frequently by both Phoenician and Greek merchants beginning in the 8th–7th centuries B.C. (Fig. 10.7). It was Aeolus, after all, the eponymous king of the Aeolian Islands, who controlled the winds in Greek mythology, including Boreas (the North Wind), which caused Odysseus so much grief on his long journey home from Troy (Fig. 10.8). Today the Malvasia Cândida varietal is grown for winemaking on several central Mediterranean islands (especially Sardinia and the Aeolian Islands; Fig. 1.7), but it is not widely used on Crete.

Unfortunately, the Malvasia di Candia wine remains an enigma. Of the three grapes that seemed to have had the best and most obvious chance of tying Malvasia wine to Crete (Malvasia Bianca di Candia, Malvasia di Candia Aromatica, and Malvasia Cândida), none could do so conclusively. But there are other interesting grape varietals that are thought to have been used by Cretan winemakers as far back as the Romans, if not before. Several of these legacy grapes (such as Athiri,

Figure 10.5. A sistrum is a musical percussion instrument associated with the worship of the ancient Egyptian goddess Hathor, and it also is seen in Minoan and Anatolian art: (a) ceramic Middle Minoan examples were excavated at Hagios Charalambos in eastern Crete, ca. 2000 B.C., height 15.8 cm, reprinted from Betancourt 2014, pl. 25B; (b) Late Minoan I bronze sistrum excavated on the island of Mochlos in eastern Crete, ca. 1525 B.C., height 28.5 cm, reprinted from Soles 2011, 135, fig. 14.2; (c) a drawing of a sistrum is featured on the label of Agelakis Winery's *Seistro* wine.

Figure 10.6. The top half of the Harvester Vase, excavated in a Late Minoan (ca. 1500 B.C.) villa at Hagia Triada in south-central Crete, shows a sistrum being used to keep the rhythm of a song or chant by a group of agricultural workers. Carved in serpentinite, this stone vessel may have originally been covered with gold leaf. Max. diam. 11.5 cm. Reprinted from Betancourt 2007, 99, fig. 5.27; courtesy P. Betancourt.

Figure 10.7. The Aeolian Islands off the northeastern coast of Sicily as they are seen from the island of Vulcano with the island of Lipari in the middle, Salina to the left, and Panarea to the right. Photo Giovanni from Catania, Sicily, https://commons.wikimedia.org/wiki/File:Eolie.jpg#/media/File:Eolie.jpg, accessed January 20, 2025; CC BY-SA 2.0, https://creativecommons.org/licenses/by-sa/2.0/.

Figure 10.8. Odysseus (holding trident) is humorously chased across the waves by the North Wind (ΒΟΡΙΑΣ' head at upper right) on a raft consisting of two (wine?) amphorae. Black-figure skyphos (drinking cup) from Thebes, fourth century B.C., height 15.4 cm. Oxford, Ashmolean Museum, acc. AN1896–1908.G.249; photo C. Raddato, https://commons.wikimedia.org/wiki/File:Boetian_black-figure_pottery_skyphos_(winecup),_found_at_Thebes,_4th_century_BC,_Odysseus_at_sea_on_a_raft_of_amphoras,_Ashmolean_Museum_(8401774652).jpg#, accessed May 21, 2025; CC BY-SA 2.0, https://creativecommons.org/licenses/by-sa/2.0/.

Liatiko, and Thrapsathiri, among others) have been rediscovered, rescued, and regrown on the island during the past few decades. Perhaps in the future one of them, or a blend involving several of them, will clarify the situation (see Stavrakaki and Stavrakakis 2017). We will look at several of these grapes in the next chapter, and, again, it may or may not be a single varietal for which we are searching. Keep tasting!

Further Reading

Betancourt, P.P. 2007. *Introduction to Aegean Art*, INSTAP Academic Press.

———. 2014. *Hagios Charalambos: A Minoan Burial Cave in Crete* I. *Excavation and Portable Objects* (*Prehistory Monographs* 47), INSTAP Academic Press.

> **The Malvasia Project**
>
> In 1997, the Tsimbidi family opened the Monemvasia Winery in Laconia with 30 hectares of organically tended vineyards and the goal of bringing back the importance of this historic vinicultural area. Today they feature wines using the Monemvasia grape (Fig. 10.9)—a varietal more at home on the island of Paros than elsewhere in the Aegean area—to produce Monemvasia, a monovarietal PGI Laconia white wine made from that grape (13% abv.), as well as Monemvasios, a dry red PGI Laconia blend (90% Agiogitika and 10% wMavroudi) with a mixed oak program (13% abv.). Perhaps more relevant here is their PDO Monemvasia-Malvasia that blends the Monemvasia grape with other island varietals (60% Monemvasia, 20% Kydonitsa, 10% Assyrtiko, and 10% Asproudi). Picked when ultra-ripe, these grapes are then dried in the sun for up to 12 days before enjoying a long rest in oak barrels (13.5 abv. with 184 g/L of R.S.). Together, these wines form an integral part of The Malvasia Project aimed at reflecting the area's past vinicultural achievements (info@monemvasiawinery.gr).

Johnson, H. 1989. *Vintage: Story of Wine*, Simon and Schuster.

Jones, H.L., trans. 1927. *Strabo: Geography*, vol. IV: books 8–9 (*Loeb Classical Library* 196), Harvard University Press.

Karakasis, Y. 2021. *The Wines of Crete: A Terroir Report*, Herakleion Chamber of Commerce & Industry, https://karakasis.mw/%cf%84he-wines-of-crete-terroir-report/, accessed May 13, 2025.

———. 2022. "The Malvasia Enigma," https://www.karakasis.mw/malvasia-enigma, accessed May 13, 2025.

Kourakou, S. 2020. *Malvasia: The Renowned Wine Yesterday and Today*, trans. A. Douma, Foinikas Publications.

Kourakou-Dragona, S. 2015. *Vine and Wine in the Ancient Greek World*, trans. M. Relaki, Foinikas Publications.

Lazarakis, K. 2018. *The Wines of Greece*, Infinite Ideas Limited.

Leonard, A. 2020. *Mediterranean Wines of Place: A Celebration of Heritage Grapes*, Lockwood Press.

Manessis, N. 2000. *The Illustrated Greek Wine Book*, Olive Press Publications.

Manolioudis, S.M. No date. *Wine Routes in the Cultural Landscapes and Malvasia of Crete*, Herakleion.

Metaxas, N.M., and H.M. Hassan. 2022. "Khandax/al-Khandaq: A New Approach to the Toponym," *Κρητικά Χρονικά* 42, pp. 167–201.

Murray, A.T., trans. 1919. *Homer: Odyssey*, vol. I: books 1–12 (*Loeb Classical Library* 104), G.E. Dimock, rev., Harvard University Press.

Robinson, J., J. Harding, and J. Vouillamoz. 2012. *Wine Grapes: A Complete Guide to 1,368 Vine Varieties, Including Their Origins and Flavours*, Ecco.

Soles, J.S. 2011. "The Mochlos Sistrum and Its Origins," in *Metallurgy: Understandy How, Learning whY, Studies in Honor of James D. Muhly (Prehistory Monographs* 29), P. Betancourt and S. Ferrence, eds., INSTAP Academic Press, pp. 133–146.

Stavrakaki, M., and M.N. Stavrakakis. 2017. *The Cretan Grapes*, Tropi Publications.

Figure 10.9. The Monemvasia grape. Photo courtesy M.N. Stavrakaki and M. Stavrakakis; reprinted from Stavrakaki and Stavrakakis 2017, 157, fig. 36.

11

The Heritage Grapes of Cretan Wine

Albert Leonard, Jr.

Although Crete has never been a stranger to those who practiced the winemaker's craft, the modern history of Cretan wine really began in the 1950s. Up to that time, most of the wine that traveled around the island moved in wineskins, wooden barrels, and large glass containers called demijohns, often clad in straw jackets to protect their contents. Then the Miliarakis brothers, who were local suppliers and bulk importers in Peza (17 km southeast of Herakleion; Fig. 1.2:24), had the novel idea of selling their wines in smaller, more personal glass bottles, complete with paper labels identifying their ΜΙΝΩΣ brand of wine (later adding two more labels, MINOIKO and *Minos Palace*; Fig. 11.1). Over the following decades they were joined by (among others) the Peza Agricultural Union Cooperative, the Lyrarakis Estate, the Alexakis Winery, Douloufakis Winery, Domaine Zacharioudakis, and, more recently, the Scalarea Estate to form the backbone of the modern Cretan wine industry. To these larger enterprises, at least forty more—superb—smaller wineries have been added over the years. Most of them are connected by the well-organized Wines of Crete program online, and it seems as if more, excellent wineries blossom on the island every year.

Recently, many Cretan wineries have begun to devote their efforts to producing a "more natural" wine. Vocabulary and criteria will vary considerably from organic (emphasizing the absence of synthetic herbicides and fertilizers as in EU Regulation 2092/91 updated by 837/2007) through biodynamic (a more holistic approach harkening back to the lectures of philosopher [and occultist] Rudolph Steiner in the 1920s; see below,

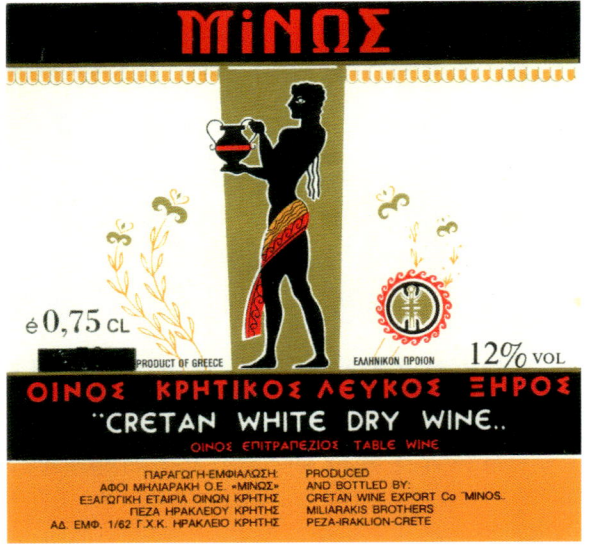

Figure 11.1. The iconic label from a 1960s bottle of *Cretan White Dry Wine* offered by Minos Winery in Peza (a town southeast of Knossos). It was the first company to offer their wine in personal-sized glass bottles (1952). The iconography is proudly based on the Procession Fresco from the Palace of Minos at Knossos (see Evans 1921–1935, II, 719–757, figs. 450–452). Also note the reverse-tapered column, the Minoan lilies, and the double axe (within the medallion at right), all Minoan motifs. Image courtesy N. Miliarakis.

Further Reading, Paull 2011). Certification showing participation in or compliance with a particular program is awarded by companies such as Demeter International in Herakleion. In the vineyard these practices can range from varietal and site selection to sustainable farming practices. In the winery itself gravity-fed juice transfer may replace mechanical pressure (and rubber hoses), or fining and filtering operations may be complete avoided. Fermentation may be left to local (as opposed to cultured or commercial) yeasts, and the addition of sulfites may be minimized. Regardless of the terminology used, every step is intended to make a better wine for you to enjoy while at the same time being considerate of the earth that provided it: "back to the future" on a beautiful island. Wineries that engage in these practices do so with a great deal of pride, and they are most willing to discuss their individual achievements with you either in person or in print.

Whether an operation is large or small, the winemakers of Crete possess a passion for the identification, protection, resurrection, and vinification of the grape varietals that are indigenous, or historically important, to their island. Over forty such grapes have been described and meticulously documented by Stavrakaki and Stavrakakis (2017) in their beautifully illustrated and very approachable volume. Over the past few decades, several wineries have rediscovered, replanted, grown, and vinified these legacy grapes from an earlier time. The sample of Cretan heritage grapes offered below (written in 2023), is not intended to be exhaustive but rather to guide your explorations. White grapes are followed by red grapes, with specific varieties presented in each category. After a short history of each varietal, the winery—cava, domaine, estate, κτήμα (estate in Greek)—is listed, followed by the name under which that wine is sold in italics, and an English translation, if noteworthy.

Often, a word or two of comment will be added that may include the level of alcohol by volume (abv.) and residual sugar (R.S.) by grams per liter (g/L) if known or thought to be either interesting or relevant. Percentages given in blends may vary slightly with the vintage and, when such information is absent, may indicate the proprietary rights of the maker. Be sure to include as many of these wines as possible in your tastings in order to fully understand the depth of the island's offerings. Be prudent but enjoy!

White Grapes of Cretan Wine

Ασυρτικό (Assyrtiko)

Assyrtiko (ah-sir-ti-KO) is a grape best known from Santorini (ancient Thera; Fig. 1.6), where it forms the backbone of that island's white wine portfolio (Fig. 11.2). Although its exact place of origin is lost to history, the grape originated somewhere in the eastern Mediterranean, and may have been brought to the Aegean by seafaring merchants from Phoenicia (modern Lebanon) more than 3,000 years ago. Today the grape is highly valued for its rare ability to survive the extreme heat of Aegean summers while retaining a crisp acidity (lime and lemon) that softens on the palate to orange and tangerine. On the island of Crete, one detects less of the salinity and minerality that defines many of the Assyrtiko wines from the more volcanic soils of Santorini, but it is there—often disguised as a slight aftertaste. Assyrtiko wines are always fresh, clean, and crisp.

Monovarietal Assyrtiko wines are crafted by several Cretan wineries including: Alexakis (*Assyrtiko*; Fig. 11.3); Diamantakis (*Petali Assyrtiko*);

Figure 11.2. The Assyrtiko grape. Photo courtesy M. Stavrakaki and M.N. Stavrakakis.

Figure 11.3. By using two Linear B ideograms on its *Assyrtiko* label, Alexakis Winery would have announced to Greek speakers at Knossos over 3,000 years ago that this was a wine that could be enjoyed by both men and women. Photo courtesy Alexakis Winery.

Digenakis (*Assyrtiko Paliarda Single Vineyard*); Domaine Economou (*Assyrtiko 2014*), a white dry wine, 13.5% abv.; Douloufakis (*Alárgo Assyrtiko White*); Lyrarakis (single area: *Vóila Assyrtiko*); Michalakis Estate (*Assyrtiko*); ΜΙΝΩΣ (*Ασύρτικο Μιλιαράκυσ*); Paterianakis (*Domaine Paterianakis Assyrtiko*); Pnevmatikakis (*2 Agrielies Assyrtiko* and *Amelliti Assyrtiko*); Strataridakis (*Assyrtiko*); and the Toplou Monastery Winery (Σαμώνιον Ασύρτικο).

Assyrtiko blends are made with several historically Cretan varietals. Diamantakis presents a fifty-fifty Vidiano blend (*Diamantopetra White*), whereas Pnevmatikakis adds 20% Assyrtiko to his Vidiano (*Eklektos*). Assyrtiko is blended with Vilana by Douloufakis (*ΕδώWhite*); and both Assyrtiko and Vidiano are added to a base of 85% Vilana by Alexakis (*Mare de Candia*), who also adds 20% Assyrtiko to a blend of both Vilana (60%) and Thrapsathiri (20%) in their *Kariki* white wine. Efrosini adds 40% Assyrtiko to Malvasia Aromatica in their *Onirikon* (Dreamlike); and the trio of Assyrtiko, Vilana, and Vidiano is also used by Mamidakis-Anoskeli in their Άνω Πλαγιά white wine.

Αθίρι (Athiri)

Athiri (ah-THIR-ee; Fig. 11.4) is an ancient Greek grape whose name suggests it has had a long tenure on the island of Thera (Santorini), where today it is used to bring a mild, lemony-citrus flavor to several of the island's blended white wines. If correctly identified as the same grape that produced the ancient *thireos oinos* (Theran wine), Athiri has been around since at least the second century A.D. A specifically Cretan Athyrin wine is mentioned in a poem attributed to the 12th-century poet known as Ptochoprodromos (Poor Prodromos), who wrote the poem during (or shortly after) the reign of Byzantine Emperor Manuel I Komnenos (r. 1143–1180). Some historians suggest that Athiri was the grape, or one of the grapes, that produced the famous Malvasia.

Alexakis' monovarietal *DandeLion* nicely shows off the subtleties of the Athiri grape, whereas the Scalarea Estate illustrates the grape's versatility in its blend of separately harvested, and individually vinified Athiri and Vidiano grapes in its *Scalarea White* wine.

Figure 11.4. The Athiri grape. Reprinted from Stavrakaki and Stavrakakis 2017, 43; courtesy M. Stavrakaki and M.N. Stavrakakis.

Δαφνί (Daphni or Dafni)

Daphni (daf-NEE; Fig. 11.5) is a thick-skinned grape that takes its name from the distinctive aroma of its must (juice), which is very reminiscent of the leaf of the bay laurel (*Laurus nobilis*), widely known for its culinary use in soups and stews. The ancient Greeks and Romans associated laurel with Apollo, the god of light and brightness, who touched almost every facet of their lives. At one point Apollo had fallen in love with a beautiful nymph named Daphne, the daughter of a river god, but unfortunately she did not share his feelings. Undaunted, Apollo continued to pursue her to the point that her very protective father was forced to turn her into a laurel tree to escape his advances. From the fragrant leaves of that tree, Apollo fashioned a wreath to remind him forever of his love for her. This story was celebrated in the first century by the Latin poet Ovid in his *Metamorphoses*, and the exact moment Daphne was transformed into a laurel tree was vividly imagined in stone by the 17th-century Italian sculptor Bernini (Fig. 11.6).

Daphni or Dafni the grape, not to be confused with the similarly sounding wine-growing region (or appellation) in central Crete (Daphnes), was brought back from near extinction by Manolis Lyrarakis in the early 1990s. He went on to nurture its comeback at the family winery in the town of Alagni (25 km southeast of Herakleion; Fig. 1.2:1). Dafni wines are characterized not only by the grape's signature aroma of laurel, but also by its combination of other dried herbs, such as rosemary and oregano. These herbs are repeated on the palate over a rich fruity

Figure 11.5. The Daphni grape. Reprinted from Stavrakaki and Stavrakakis 2017, 55; courtesy M. Stavrakaki and M.N. Stavrakakis.

Figure 11.6. Marble sculpture of Apollo and Daphne by Italian Baroque artist Gian Lorenzo Bernini. Ca. 1622–1625, height 2.43 m. Photo Architas, https://commons.wikimedia.org/wiki/Category:Apollo_and_Daphne_(Bernini)#/media/File:Apollo_and_Daphne_(Bernini)_(cropped).jpg, accessed May 22, 2025; CC BY-SA 4.0, https://creativecommons.org/licenses/by-sa/4.0/.

base with good acidity and relatively moderate, often 12%–12.5% (abv.), alcohol levels.

Monovarietal Daphni wines include Lyrarakis Estate's *Psarades Dafni* wine crafted from sequentially harvested and separately vinified grapes from its Psarades vineyard near Alagni, 12.5% (abv.), and Michalakis Estate's cold fermented Δαφνί, at 12% (abv.).

Daphni also blends well in trials with Cretan heritage grapes such as Muscat, Plyto, Thrapsathiri, and Vilana, as well as with international varietals such as Digenakis' half and half blending of Dafni with Sauvignon Blanc in its *Αερικό* dry white wine (13% abv.).

Μοσχάτο Σπίνα(ς) (Muscat Spiná[s])

The name Muscat Spiná(s) (mus-kat spee-NA[S]; Fig. 11.7), according to scholars (Robinson, Harding, and Vouillamoz 2012; Stavrakaki and Stavrakakis 2017), is a synonym for Muscat Blanc à Petits Grains, a varietal that is also known as Moscato Bianco (White Muscat) and Moschato Samou (Muscat of the Island of Samos). The grape is commonly thought to have originated in Greece where it appears on wine labels in a confusing variety of forms: Spinas Muscat, Moschato Spina, Moschato Spinas/Mazas, etc. (with or without the terminal "s"), especially in the western part of Crete. Long considered a rare, thin-skinned variety of the muscat grape, it was recognized by Yiannis Konstantakis of Scalarea Estate and brought to Herakleion where it is now cultivated and bottled (as Φανταξομέτοχο wine). Because of the grape's sweetness, some historians would equate it with the varietal that first-century historian Pliny the Elder identified as Melissa (the grape of the bees) and would suggest that the grape had been imported to western Crete before he lost his life in the eruption of Mt. Vesuvius (A.D. 79). Since the mid 18th century, the grape has been associated closely with the Cretan village of Spiná, near Chania (Fig. 1.2:29).

Monovarietal Muscat Spiná wines include Agelakis Winery's *Materia* and Alexakis Winery's Μοσχάτο Σπίνας. Boutari has released *Moschato Spinas*, a sweet (hand-picked and sun-dried seven days) version that rests for six months in old oak barrels (14% abv.). Also included in this group are: Digenakis' *Marisini White* (13% abv.); Dourakis Winery's *Kudos*, Greek for high achievement (13.3% abv.); Domaine Gavalas's *Fragospito*,

Figure 11.7. The Muscat Spiná(s) grape. Reprinted from Stavrakaki and Stavrakakis 2017, 94, fig. 27; courtesy M. Stavrakaki and M.N. Stavrakakis.

a blend of *Malvazia* and *Moshato Spinas* from their eponymous Fragospito vineyard (13% abv.) named after the old Venetian residence in the middle of their vineyard at Vorias Monofatsiou; and Manousakis Winery's *Nostos* (meaning the journey). Michalakis Estate presents *Moschato Spinas* both as a dry wine (12.5% abv.) and as sparkling (12.5 abv.). Domaine Paterianakis' *Moschato Spinas* (12% abv.), Pnevmatikakis' *Sweet Magic Red* (12.5% abv.), and Strataridakis' *Castelanos* (a sweet white Moschato) are wines produced from this grape.

Michalakis Estate blends separately vinified Moschato with Mandilari in its medium sweet M^2 *Rosé Sto Tetragono* (11.5% abv.) and adds Moschofilero to its Moschato in its medium sweet M^2 *White Sto Tetragono* (11% abv). Pnevmatikakis Winery combines Moschato with Romeiko in their *Vin de Crete* dry white wine, and blends Moschato with Vilana in its medium sweet Όνειρο (Dream).

Πλυτό (Plyto)

Plyto (plee-TO; Fig. 11.8) is a grape that has been grown on Crete since at least the 14th century and may very well be indigenous to the eastern third of the island. It was there that the grape was identified and rescued by Manolis and Sotiris Lyrarakis in the 1990s (Fig. 11.9). The grape is also grown on the neighboring island of Kythera, where it is known as Ploto. White peach, edgy citrus, and green Anjou pear fluctuate on the palate above a sub-stratum of minerality.

Figure 11.8. The Plyto grape. Reprinted from Stavrakaki and Stavrakakis 2017, 107; courtesy M. Stavrakaki and M.N. Stavrakakis.

Figure 11.9. The rescued Plyto grape is now enjoying life and thriving in its new home at the Lyrarakis Winery. Photo A. Leonard.

Monovarietal Plyto wines are relatively scarce due to the limited availability of the fruit. Lyrarakis offers a single vineyard *Plytó Psarades* wine with grapes picked at two stages of ripeness (12.8% abv.). Pnevmatikakis Winery makes *Plytó Open Book* from grapes grown around Kissamos (Hagia Irene parish) at the western end of the island, and Michalakis Winery presents the grape in its Estate Series simply as *Plyto* (13% abv.).

Plyto blends are even scarcer, but they exist. The Rhous-Tamiolakis Winery in Peza adds 30% Plyto to its Vidiano to produce the dry white wine they market as *Skipper White*, whereas Silva-Daskalaki adds a small amount to a biodynamic and organically grown Sauvignon Blanc base to produce its *Sera White* that spends 16 months in French oak barrels (12% abv.).

Θραπσαθίρι (Thrapsathiri)

Thrapsathiri (thrap-sa-THIR-ri; Fig. 11.10) is a native Cretan grape, but the once-held view that its name had been derived from an amalgamation of the Greek word *thrapsa* (lots of) and that of the Athiri grape has recently been disproven by DNA analysis (Stavrakaki and Stavrakakis 2017). Thrapsathiri is the same as the grape known as Begleri that is grown in southern Greece and the Cycladic islands. It may have been brought to Crete under that name, where it subsequently assumed the name Thrapsathiri. If the red-berried Begleri Kokkino is a color mutation of Thrapsathiri, that fact may be relevant to discussions on the color of the historic Malvasia/Monemvasia wine. The Thrapsathiri grape is also thought to share a close relationship with the Cretan varietal Vidiano because all three of these varietals have been proposed at one time or another to have been components of the enigmatic Malvasia (see below, Further Reading, Stavrakaki and Stavrakakis 2017).

A Thrapsathiri wine opens with a faintly floral fragrance of ripening peach and cantaloupe, then morphs into that of riper fruits (even pineapple) on the palate, where it develops into a balanced, softly acidic, and well-structured wine.

Monovarietal Thrapsathiri wines include two from Domaine Economou: *Discovery*, from 70-year-old ungrafted vines in eastern Crete (13.5% abv.), and *Odyssey 2012*, a dry white wine from grapes from old rootstock vines fermented with indigenous yeasts (13.5% abv.). Also on the list are Idaia Winery's *Ocean* (12.5% abv.), Toplou Monastery Winery's *Γερτό* (Bent), named after the wind-bent trees near its hill-top vineyard, and

Figure 11.10. The Thrapsathiri grape. Photo courtesy M. Stavrakaki and M.N. Stavrakakis.

Lyrarakis Winery's *Armi Thrapsathiri*, a wine that has spent time in both acacia and French oak barrels (13.2% abv.).

Thrapsathiri is also a popular blending grape; Miliarakis Winery adds 20% Muscat Spina to Thrapsathiri in his *2 Pharaggia* (2 Gorges), whereas Vilana (60%) is added to Thrapsathiri (40%) by Domaine Economou for both *Heliades 2014* (Daughters of the Sun), as well as *Sitia White V.Q.P.R.D. 2015*. The grapes for both wines also come from old rootstock vines that have been fermented with indigenous yeasts (13.5% abv.). Strataridakis Winery's *Strata* is a blend of these same two grape varietals (12.5% abv.), as is Toplou Monastery Winery's *Thrapsathiri-Vilana*. This appears to be a trend that was noticed early (Robinson, Harding, and Vouillamoz 2012). Stylianos Winery adds both Vilana and Vidiano to a Thrapsathiri base for their *Theon Dora* (Gift of the Gods). The grape's ability to work well with international varietals can be seen at Efrosini Winery, where Thrapsathiri (40%) is added to their Chardonnay to produce the oak-aged *Lumicino* blend (13.0% abv.), as well as at Paterianakis Winery, which adds 30% Sauvignon Blanc to a Thrapsathiri base to achieve its *Melissinos* white wine (12.5% abv.).

Μελισάκι (Melissaki)

Melissaki (me-lih-SAHK-ee; Fig. 11.11) is another Cretan legacy grape that was recognized, rescued, and replanted by the Lyrarakis family at their eponymous winery. Considered to be "unicorn rare" by Konstantinos Lazarakis (the first Greek Master of Wine), the grape's name Melissaki shows an undefined connection with the Greek word for honey (μέλι) or honeybee (*Apis mellifera*); it, however, has been proven not to be related to the ancient *Vitis apiana* (vine of the bees), whose sweet wine (*Apianum*) delighted both the Roman Columella and Pliny the Elder. Transplanted from the foothills of Mt. Psiloritis (in central Crete; Fig. 1.2:23) to the family owned, non-irrigated Gero Detis vineyard in Alagni (Fig. 1.2:1), a monovarietal *Melissaki Gerodeti* wine has been offered annually since 2016. Handpicked and spontaneously fermented on vineyard-specific yeasts (Bio certificate), it is naturally stabilized after a rest in oak before it is bottled (unfiltered and unfined) with a minimum (40 ppm) shot of sulphur dioxide. Golden yellow, tending toward orange in certain light, it presents honey, dusky herbs and citrus zest on the nose, followed by candied citrus on the palate with slightly tannic undertones toward the finish. The 2021 wine was released at 13.5% abv.

Figure 11.11. The Melissaki grape. Photo courtesy M. Stavrakaki and M.N. Stavrakakis.

Βιδιανό (Vidiano)

Vidiano (vid-i-a-NO; Fig. 11.12), often called the Greek Viognier and referred to as the White Diva of Cretan wine (Klados Winery), is a grape that most probably originated in the region of Amari, to the south of Rethymnon in central Crete (Fig. 1.2:27). Genetically linked to both Thrapsathiri and Vilana, its very existence was threatened until Minas Tamiolakis and Nikos Douloufakis identified it and replanted it in their respective vineyards (Tamiolakis Winery has subsequently changed its name to Rhous Winery). Vidiano wines are golden in color, often with a slight greenish tinge that predicts citrus and floral elements on the nose, as well as hints of white pepper that combine to announce a full-bodied wine with a long finish and slightly mineral aftertaste.

The grape is very versatile both in the vineyard and the winery. Domaine Economou offers *Vidiano* wine made from fruit from ungrafted 40-year-old vines. Endochora Winery, in the region of Chania, has grafted its Vidiano stock onto old Romeikos roots rather than selecting one of the more popular North American root systems (Fig. 11.13 for Endochora's *Vidiano* label with the Cupbearer from the Cycladic Islands). Lyrarakis makes three monovarietal wines from the grape including *Minimus*, whose fruit is picked in the cool of the night when its aroma is high and its sugar is low. Lyrarakis also produces a diet-friendly wine (9.3% abv.) that comes in a modest 375 ml bottle. Silva-Daskalakis Winery ferments the grape on indigenous yeasts in 300-liter clay pitharia (storage jars). The wine rests in oak barrels for three months and is bottled unfiltered and without sulfites with a sealed recyclable stopper.

Monovarietal Vidiano wines are popular. They include Diamantakis Winery's *Vorina Vineyard Vidiano* (13.3% abv.); Digenakis Winery's *Βίος Πρώτος* (Prime Life); Dourakis' *Λύχνος* (Oil Lamp; 12% abv.); Domaine Gavalas's *Vidiano Gavalas* (13% abv.); Efrosini Winery's *Mikri Evgeniki Vidiano*; and Idaia Winery's *Vidiano* (13.5% abv.). Before the forest fire that destroyed

Figure 11.12. The Vidiano grape. Photo courtesy M. Stavrakaki and M.N. Stavrakakis.

Figure 11.13. Stressing the relationship between wine and culture, the Endochora Winery uses the image of an early "cupbearer" from the Cycladic Islands on its labels. The original piece—which dates to the Early Cycladic II period, ca. 2700–2400/2300 B.C.—is in the Goulandris Museum of Cycladic Art, Athens, acc. ΝΓ0286. Image courtesy M. Tsafarakis.

so much of her vineyards in 2022, Iliana Malihin produced two monovarietal Vidiano wines at her winery in Melampes (Rethymnon region) from minimally handled fruit from vines widely separated in age: *Amygdalos Old Vines Vidiano* from hand-harvested grapes from vines said to be around 120 years old, unfiltered and unfined (14% abv.), and a *Young Vines Vidiano* wine from newer, 15-year-old vines (at 13% abv.). Lyrarakis offers two: *Vidiano* (12.4% abv.), as well as an oak-aged Vidiano from their *Ippodromos Vineyard* (14.4%). Domaine Paterianakis also offers two: *3.14 Vidiano* and *Melissokipos Vidiano*. Rounding out the list are Karavitakis Winery's *Klima Vidiano* (13% abv.); Klados Winery's *Aloides Vineyard White Diva* (13% abv.); Manousakis Winery's *Nostos Vidiano* (14.4% abv.); Michalakis Estate's *Vidiano* (12.5% abv.); Miliarakis' *Vidiano*; Pnevmatikakis' *Vynos* (12%; abv.); Strataridakis' *Άσπρα Χαράκια* (White Rocks; 13.5% abv.); and Domaine Zacharioudakis' *Vidiano* (14% abv.).

Vidiano blends well with both Greek and international varietals. Scalarea Estate's *Scalarea White* is a blend of separately harvested and individually vinified Vidiano and Athiri grapes. The Rhous-Tamiolakis Winery in Peza adds 30% Plyto to its Vidiano to produce the dry white wine bottled as *Skipper White*; Pnevmatikakis adds Vidiano to Romeiko in its *Vin de Chania* white wine (11.5% abv.); Lyrarakis separately vinifies 15% Vidiano and adds it to its Muscat Spina to produce its *Lyrarakis White* (12.75% abv.); whereas Klados Winery adds 40% Vidiano to its Muscat Spina to create what it calls *Great Hawk*. Michalakis Estate separately vinifies Vidiano, Muscat, and Chardonnay before combining them into their *Gold Cuvee White* (13% abv.). In 2019, Lyrarakis Winery added 2% Vidiano to 77% Syrah and 21% Mandilari to tweak the flavors of its *Symbolo Grand Cuvee* (14.4% abv.), whose name commemorates the symbol for wine (Fig. 11.14) that appears on clay tablets over 3,000 years ago in the earliest known form of the written Greek language (ideogram AB 131 in Linear B; see Fig. 5.3). A similar sign is known from the earlier, pre-Greek "Minoan" language of Crete (Linear A), and it may ultimately be related to an even earlier Egyptian Hieroglyph (see Ch. 5).

Figure 11.14. The label for Lyrarakis Winery's *Symbolo* wine commemorates the symbol for wine (to the right of the name) used over 3,000 years ago on clay tablets in the earliest-known form of the written Greek language, Linear B (wine symbol known as ideogram AB 131). Image courtesy K. Lyrarakis.

Βιλάνα (Vilana)

Vilana (vil-LA-na; Fig. 11.15) is the rock star of Cretan white wine production, a pale-yellow cultivar that produces fresh, fruity, and aromatic wines in a wide variety of styles. The green apple crispness taste, combined with a relatively low level of alcohol (often ca. 11.5–12%), make these wines especially food friendly, and best when consumed young. Chances are, order white wine in Herakleion and it will most probably contain some percentage of Vilana. Single varietal PDO (Protected Designation of Origin) and PGI (Protected Geographical Indication) Vilana wines are produced by many Cretan winemakers.

Monovarietal wines include Boutari's very popular *Kretikos* Vilana (11.5% abv.); Anoskeli-Mamidakis' *Anoferia*; Dourakis Winery's *Monoceros Vilana* (12.9% abv.); Gavalas's *Vilana*; and Idaia Winery's *Ideia GI Vilana* (13% abv.). Lyrarakis Winery offers a *Vilana* (12.5% abv.) and a *Vilana Pirovolikes* (13% abv.) that has spent some time in acacia as well as French and American oak barrels. Miliarakis, the first Cretan winery to age its white wines in oak, features a *Vilana Fumé*, which bends the flavor curve toward apricot/peach while picking up a layer of vanilla from time spent in the barrel, in addition to their more standard *Minoa Palace* and *35° North 25° East White*.

Vilana blends successfully with other Hellenic grape varieties. Agelakis Winery uses Malvasia Aromatica in its *Seistro* (Sistrum), and Michalakis blends it with Vidiano in their white *Vin de Crete White* (11.5%. abv.), whose label features a rare, mid-fifth century B.C. Knossian silver *stater* (coin) depicting a running Minotaur (Fig. 11.16). The juice of both

Figure 11.15. The Vilana grape. Reprinted from Stavrakaki and Stavrakakis 2017, 138; courtesy M. Stavrakaki and M.N. Stavrakakis.

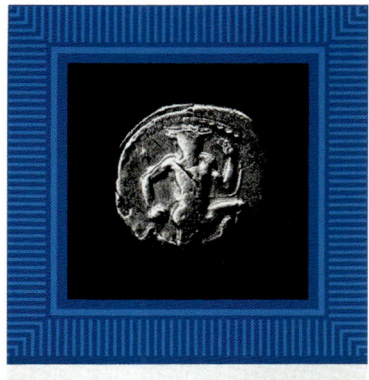

Figure 11.16. Michalakis Winery pays homage to the island's Minoan past by featuring the image of a rare silver stater (coin) from mid-fifth century B.C. Knossos that depicts a running Minotaur on its *Vin de Crete* Vilana/Vidiano blend. Image courtesy Michalakis Winery.

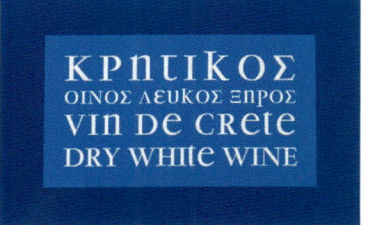

Vidiano and Muscat Spina are blended by Digenakis into their medium sweet *Kernos* wine; whereas Michalakis blends three grapes (Vidiano, Moschato, and Moscofilero) to achieve their *Merastri White* wine.

Red Grapes of Cretan Wine

This section emphasizes six local, or historical, Cretan varieties of red wine: Kotsifali, Ladikino, Liatiko, Mandilari(a), Melissaki, and Romeiko.

Κοτσιφάλι (Kotsifali)

Kotsifali (kot-si-FAH-li; Figs. 11.17, 11.18) is associated with the modern Cretan appellations of Peza and Archanes (regions southeast of Herakleion; Fig. 1.2). A low-acid grape with a deficiency of anthocyanins, it yields weak red wines (often with an orange tinge), but its naturally high sugar levels offer greater alcohol and longer aging potential. Varietal Kotsifali red wines are less common than blends, although the grape's rich flavors of red fruit and spice come through in each of the very different pair of wines that are made by Lyrarakis: the straightforward *Kotsifali* (13% abv.) and the single-area *Karnari Kotsifali*, vinified in the rustic, west Cretan Marouvas style, with spontaneous fermentation and varying degrees of oxidation, before enjoying two and a half years in previously filled French oak barrels (13% abv.). Miliarakis Winery even makes a still, white wine (*blanc de noir*) from this red grape.

Figure 11.17. The Kotsifali grape. Reprinted from Stavrakaki and Stavrakakis 2017, 75; courtesy M. Stavrakaki and M.N. Stavrakakis.

Figure 11.18. Endochora Winery takes great pride in its 2021 monovarietal Kotsifali wine with a distinctive label that would stand out on any shelf or table. Image courtesy M. Tsafarakis; see also Fig. 11.13.

Monovarietal Kotsifali Wines include Domaine Paterianaki's *Kotsifali 3.14*; Endochora Wines' *Kotsifali* (Fig. 11.18); Idaia Winery's *Fresco Rosso* (13% abv.); and Klados Winery's *Pink Blackbird* rosé (13% abv.). Michalakis produces an *Estate Kotsifali* (12.5% abv.); Pnevmatikakis offers three choices: two rosés, *Pole of Attraction* and Pink Panther, as well as *Alovitos*, a dry red from Hagia Triada parish (in western Crete). Strataridakis' contribution is *Strata Kotsifali* (12.5% abv.), whereas Silva-Daskalaki's *Grifos Rosé* is a biodynamic/organic dry wine made from sun-dried Kotsifali grapes that have been fermented in 300-liter clay pitharia (storage jars) using native yeasts. They are transfered (unfiltered) in 500 ml glass bottles with a recyclable cane stopper (14.5% abv.). Zacharioudakis Winery completes the meal with *Epilogue*, a sweet dessert wine extensively aged in oak barrels (14% abv.).

Kotsifali/Mandilari blends are numerous. Because Kotsifali's weakness is countered by the Mandilari grape's strength (see Mandilari below), they work very well together and form the backbone of the Cretan red wine repertoire. A ratio of 70% Kotsifali to 30% Mandilari is a good starting point for blending many of the Peza/Archanes red wines. Vinification methods vary considerably and may even include a little time in oak barrels. A good introduction to Kotsifali/Mandilari blends is the widely available Boutari's *Kretikos Red* (Fig. 11.19) or the *Scalarea Red* (14% abv.). Kotsifali is usually the predominant varietal in such blends, but Alexakis Winery's *Kariki* reverses the norm and uses Mandilari (60%) as the major component, thereby leveling the acids and tannins. Efrosini Winery's *Onirikon* (Dreamlike) treats the two grapes equally (50% each) and gives the resulting blend a little time in oak barrels.

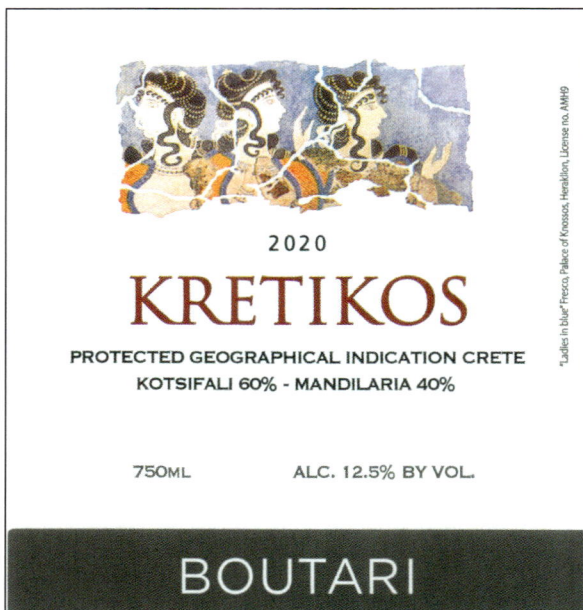

Figure 11.19. The fresco of the Ladies in Blue from Knossos adorns the label of Boutari's *Kretikos* red wine (for the fresco, see Evans 1921–1935, I, 544–547, fig. 397). Image courtesy Boutari Wines of Crete.

Idaia Winery's *Idaia GI*, spends eight months in oak (13.5% abv.). Michalakis Estate blends Kotsifali and Mandilari in both its *Vin de Crete Rosé* (11.5% abv.) and its *Vin de Crete Red* (12% abv.), whose label also bears the image of the 2,500-year-old Minotaur coin from Knossos (see Fig. 11.16). Miliarakis Winery employs a high ratio of Kotsifali (80%) to Mandilari (20%) in its *Minos Miliarakis* red blend, but uses different percentages of these grapes in its entry-level *Minos Palace Red*. The same two grapes are used equally in their *Imiglykos* (medium sweet) red blend, and different percentages of the two grapes are used in *35° North 25° East Red* and *35° North 25° East Rosé*, both of which have spent some time in oak. Paterianakis presents two such blends, *Melissokipos Kotsifali-Mandilari*, a simple half-and-half dry red, and *Domaine Paterianakis*, an unfiltered blend using separately vinified Kotsifali (80%) and Mandilari (20%) grapes. Rhous-Tamiolakis' *Skipper Red* is a blend of 70% Kotsifali and 30% Mandilari.

Λαδικινό (Ladikino)

Ladikino (lad-ik-i-NO; Fig. 11.20) is a red grape whose name is derived from the city of *Laodicea ad mare* (near the modern town of Latakia on the Syrian coast; Fig. 1.7) whose origins go back to the time of the Phoenicians. The Latakia province has been a wine producing area for centuries. Today, on Crete, the juice of this grape is used as a blending agent to increase the aromatic qualities of certain wines, especially in the Lasithi area, but none to my knowledge has been bottled commercially.

Figure 11.20. The Ladikino grape. Reprinted from Stavrakaki and Stavrakakis 2017, 78; courtesy M. Stavrakaki and M.N. Stavrakakis.

Λιατικό (Liatiko)

Liatiko (lee-ya-ti-KO; Fig. 11.21) is a vigorous, early ripening red grape that produces its clusters of small, oval berries by mid-summer, as its name (an abbreviation of *louliatiko*, meaning of July) would imply. When fermented dry, Liatiko makes high alcohol, low acid wines with soft tannins that improve with some age. Especially well made is Douloufakis' single varietal, PDO *Dafnios* from vineyards in the village of Daphnes. Its aromatics and fruity flavors led the 2012 vintage to be described as pretty sexy in the *Wine Advocate* (December 20, 2012; see also Leonard 2020, 37). On the (naturally) sweeter side, Douloufakis' *Helios* is a dessert wine made from grapes dried in the sun for about seven days, which, after a classic white vinification process, rests in French oak for ten years before being sold in 500 ml bottles (13.6% abv. with 125 g/L of R.S.). Domaine Economou's *Sitia Red V.L.Q.P.R.D. 2006*—a red *vin naturellement doux* (naturally sweet wine) from small-berried fruit grown on ungrafted, ca. 70-year-old vines—was released recently on the owner/winemaker's schedule (15% abv.). With wines such as these, and those listed below, it is no wonder many believe that Liatiko was the grape (or at least one of the grapes) used to make Malvasia, the sweet wine that has brought fame to the island since the 15th century.

Monovarietal Liatiko wines include Scalarea Estate's Λουλιάτικο, a sweet wine made from sun-dried grapes (spread on nets for 10–12 days), which, after vinification, are aged in oak barrels for three years and then transferred to 500 ml bottles when they reach 140 g/L of R.S. (15% abv.).

Diamantakis Winery produces *Petali Liatiko* (13.1% abv.), whereas Efrosini Winery offers two choices: *Mikri Evgeniki* (14.5% abv.) and *Aureo*, a naturally sweet wine from grapes that are sun-dried for 8–10 days, stored for 36 months in used oak barrels (2–3 fills), and, then, transferred to 500 ml glass bottles (13.5% abv.). Idaia Winery also offers two choices: *Potamida Liatiko*, a dry red PDO Daphnes wine (15% abv.) and *Liatiko*, a sweet red wine made from grapes that have been sun-dried for 8 days and then stored in oak barrels for three years (12.5% abv.). Lyrarakis offers a choice of three monovarietal Liatiko wines: *Lyrarakis Liatiko* (Fig. 11.22 with figures of ladies, possibly goddesses, from Crete's Geometric period, ca. 900–700 B.C.), a dry red wine from hand-harvested grapes from non-irrigated vines (13.5% abv.), *Liatikos Kedros Rosé*, a dry rosé wine from a

Figure 11.21. The Liatiko grape. Reprinted from Stavrakaki and Stavrakakis 2017, 83; courtesy M. Stavrakaki and M.N. Stavrakakis.

non-irrigated, pre-phylloxera vineyard on Mt. Kedros (13.2% abv.), and *Liatiko Aggelis* using grapes sourced from a vineyard in Siteia that was planted in the 1930s on native rootstock (13.5% abv.). Silva-Daskalaki crafts its dry *Grifos Red* by adding local yeast in 300-liter ceramic pitharia (storage jars) with a four-month rest in oak barrels before being bottled unfiltered. It also makes a second, sun-dried (6-8 days) monovarietal Liatiko, *EMILIA*, with an extended maceration process and longer aging in French and American oak barrels, before being bottled in 500 ml glass. Toplou Monastery Winery (Siteia; Fig. 1.2:28) presents two organic options: the sweet *Linós Liatikos*, named after the early 18th-century winepress (*lihnós*) at the monastery, and a rosé, *Amaranton Liatikos* (The Unfading Rose), an homage to one of the monastery's proudest possessions: an ikon of the Virgin Mary, the Ρόδον τό αμάραντον (or Unfading Rose), painted by Stamatios of Crete in 1771.

Liatiko blends are less common than monovarietals. Domaine Economou's *Mirabello 2015* (named after the Gulf of Mirabello in eastern Crete; Fig. 1.2) is a dry red blend of Liatiko (60%) and Mandilari (40%) grapes from old rootstock vines fermented with indigenous yeasts (13.5% abv.). Idaia Winery offers an incredibly complex rosé, *Venerata Rosa*, a

Figure 11.22. Labels from two of Lyrarakis Winery's 2021 monovarietal releases portray what appears to be a pair of very well-dressed women (goddesses?) from Crete's Geometric period (ca. 900–700 B.C.). One woman holds a pair of birds (left, the Vidiano wine) while the other holds stylized Aegean flowers (right, the Liatiko wine). Photo A. Leonard.

proprietary blend combining Liatiko, Muscat Spina, and Kotsifali (13.5% abv.); whereas Pnevmatikakis Winery has released *Spirit,* a medium sweet blend of Liatiko and the international grape, Syrah (12.5% abv.). Finally, Klados Winery offers *Red Nest*, a dry red blend of Liatiko (30%) and Cabernet Sauvignon (70%) that spends little time in oak barrels (13.3% abv.).

Μανδιλάρι(α) (Mandilari[a])

Mandilari(a) (man-di-LAR-i-ya; Fig. 11.23), whose name appears to come from the Greek word μαντήλι (handkerchief), are grapes that are indigenous to the island of Crete. This variety grows particularly well in the Peza and Archanes appellations (approximately 20 km southeast of Herakleion; Fig. 1.2), where the late-ripening fruit produces wines of a deep, ruby color. Although these wines are aromatic and exhibit rich, dark stone fruit on the palate with acceptable tannins, they tend to be low in alcohol.

Monovarietal Mandilari wines include Idaia Winery's *Hestia Mandilari*, which spends a year in oak barrels (13% abv.), and two from Lyrarakis Winery, which crafts both a basic *Mandilari Rosé* (12.5% abv.) and a dry red wine from its *Plakoura Vineyard*, in which ca. 12% of its hand-harvested grapes have been sun-dried for three days, vinified, and then treated to a varied program of French and American oak (14.2% abv.). Michalakis Estate also produces two monovarietals: a *Mandilari Rosé* (12.0% abv.) as well as a dry red wine with less time in oak (12.6% abv.). Stylianou Winery offers *Great Mother*, an organic, unfiltered, and unfined dry red wine (13.7% abv.; Figs. 11.24, 11.25). Most of the more standard Mandilaria/Kotsifali blends are listed above with the Kotsifali blends.

Related Mandilaria and/or Kotsifali blended wines present quite a variety: Diamantakis' *Diamantopetra Red* is a blend of separately harvested and individually vinified Mandilari (30%) and Syrah (70%), followed by a complex French and American oak program (12.5% abv.). Domaine Gavalas uses a proprietary blend of Kotsifali and Mandilari in its *Efivos* (Ephebe) rosé wine and adds Syrah as a third grape to its *Efivos* dry red wine. Douloufakis Winery uses 60% Kotsifali and a proprietary blend of Mandilaria and Liatiko in its *Enotria Rosé* (13.3 abv.); Efrosini Winery's organic, dry red *Tenebrae* adds 30% Kotsifali to 70% Syrah before giving the blend a year in French oak barrels. They also use the same triad of grapes (40% Kotsifali, 40%

Figure 11.23. The Mandilari(a) grape. Reprinted from Stavrakaki and Stavrakakis 2017, 87; courtesy M. Stavrakaki and M.N. Stavrakakis.

Figure 11.24. The label on Stylianou's monovarietal Mandilari wine, *Great Mother*, features a woman holding clusters of grapes in imitation of the reptiles that were brandished by the famous Minoan Snake Goddess from Knossos (see Evans, 1921–1935, I, 500–510, frontispiece, figs. 359–362). The winery's label identifies her as the Great Mother, using signs from the ancient Linear B script: *ma-te-re* at left and *te-i-ja* at right. Photo A. Leonard.

Figure 11.25. The famous Snake Goddess figurine made of faience was excavated by Arthur Evans in 1903 at the site of Knossos. To many, it has become an icon for the Minoan culture as a whole. Ca. 1600–1500 B.C., height 34.2 cm. Herakleion Archaeological Museum. After Betancourt 2007, 96, fig. 5.24.

Mandilari, and 20% Liatiko) in their *Demi Rosé* (12.5% abv.). Michalakis Estate's *Thèse 11 Limited Edition* blends hand-harvested Kotsifali and Syrah grapes (12% abv.). Scalarea Estate uses a half and half mix of estate-grown fruit that is separately vinified before spending 12 months in American oak. Strataridakis' 50/50 blend of Kotsifali and Syrah in Ιχνηλάτης (Harehound) vinifies hand-harvested fruit that matures in French oak (half new and half second use), before aging in glass for one year before it is released.

Ρωμέικο (Romeiko)

Romeiko (ro-may-i-KO; Fig. 11.26) is a grape that is thought to have come to Crete from Italy, and its name may derive from ρώμι (*romi*), the Greek word for strength or vigor, which would certainly be an apt description of the vine (see Stavrakaki and Stavrakakis 2017). It is a complex grape whose ancestry is shared by three genotypes, resulting in the possibility of having three different-colored berries present on the same bunch (cluster). The grape is at home around Kissamos (Chania) in western Crete, where it is used to make a variety of wines that are high in alcohol and quick to oxidize (such as sherry and Madeira wines). Traditionally, Romeiko has been used to make a wine in the style called

Marouvas ("aged" in the local dialect), especially geared to home consumption. Grapes would be harvested at or around the degree of sugar concentration of 16 degrees Baumé (1 Baumé equals 1.8 Brix, equals 18 g/L of sugar, which should result in 1% alcohol). Brix is a measure of the dissolved solids in a liquid, commonly used to measure the amount of dissolved sugar in a solution. After being partially fermented, Marouvas would be sealed (and buried!) for several years until it was opened to celebrate an important milestone in a person's or family's life. It is also customary to serve this wine at the end of a meal, often with a slice or two of apple.

Monovarietal Romeiko wines exist. Dourakis Winery produces four such wines, each with its own personality: *Dourakis Apus Romeiko*, described by the winery as "a yellow wine from a red grape" (12.3% abv.); *Dourakis Romeiko Diaselos*, termed an *orange de noir* (orange from black [grapes]), again noting that it is an orange-colored wine from a red grape that had spent 24 months in oak (13.5% abv.); *Dourakis Romeiko Cassiopeia* (named after the constellation), a brut sparkling wine produced by the *méthode traditionnelle* (traditional method) with 6.3 g/L of R.S. (12.2% abv.); and *Dourakis Euphoria* with grapes dried in the sun for up to two weeks, with 120 g/L of R.S. (11% abv.). Endochora Winery produces a light colored *Romeiko Rosé* (13.6% abv.). Karavitakis Winery produces a sweet red *Liastos* wine from Romeiko grapes that are picked when they are ultra-ripe and dried in the sun for around 12 days, then aged three years in oak barrels (15% abv.). Manousakis Winery offers *Nostos Romeiko* (the Journey Home; 14% abv.), while Pateromichelakis Winery produces its *RO Red* (Rho-Omega), a BIO Romeiko *blanc de noir* wine (white from black [grapes]; 13% abv.). Pnevmatikakis Winery offers three wines from the Romeiko grape, ranging from *Magani*, an early harvested *blanc de noir*, through Ηλίαστος *Romeiko*, a red sweet wine from grapes sun-dried for 10 days, to the more traditional Μηθύνειος that they describe as a Cretan Marouvas wine (14% abv.).

Romeiko blends include (at least) three by the Manousakis Winery (Chania): the light rosé *Nostos Pink 2020* achieved by adding 40% Grenache Rouge and 20% Syrah to its Romeiko base (13.5% abv.), the dry red *Manousakis MRS*, a blend of 30% Romeiko and 70% Syrah (14% abv.), and one done in collaboration with the Karanika Winery (northwest Greece) to produce a sparkling wine from a 50/50 blend of Romeiko from Chania and Xinomavro from Florina. The wine

Figure 11.26. The Romeiko grape. Photo courtesy M. Stavrakaki and M.N. Stavrakakis.

is branded with the name (and face) of *Hartman Molavi,* a mysterious composite character thought to epitomize the blend itself, *blanc de noir, méthode traditionnelle* (11.5% abv.).

Further Reading

Betancourt, P.P. 2007. *Introduction to Aegean Art*, INSTAP Academic Press.

Brady, M. 2021. *The Cretan Wine Guide: A-Ω in Cretan Wine. A Guide to Cretan Indigenous Grape Varieties and Wineries*, independently published.

Evans, A.J. 1921–1935. *The Palace of Minos at Knossos* I–IV, Macmillan and Co., Limited.

Karakasis, Y. 2021. *The Wines of Crete: A Terroir Report*, Herakleion Chamber of Commerce & Industry, https://karakasis.mw/%cf%84he-wines-of-crete-terroir-report/, accessed May 13, 2025.

Lazarakis, K. 2005. *The Wines of Greece*, Mitchell Beazley.

Leonard, A. 2020. *Mediterranean Wines of Place: A Celebration of Heritage Grapes*, Lockwood Press.

Manessis, N. 2000. *The Illustrated Greek Wine Book*, Olive Press Publications.

Paull, J. 2011. "The Secrets of Koberwitz: The Diffusion of Rudolf Steiner's Agriculture Course and the Founding of Biodynamic Agriculture," *Journal of Social Research & Policy* 2 (1), pp. 19–29.

Robinson, J., J. Harding, and J. Vouillamoz. 2012. *Wine Grapes: A Complete Guide to 1,368 Vine Varieties, Including Their Origins and Flavours*, Ecco.

Selinger, H. 2023. "What's the Difference between Organic and Biodynamic Wine?" *Wine Enthusiast*, https://www.wineenthusiast.com/basics/whats-the-difference-between-organic-and-biodynamic-wine/?srsltid=AfmBOorngVsbxTsS1z4B50-WZvFvUAJxuIF1HyUFWZaKe5eHrLCMA2kS, accessed May 13, 2025.

Stavrakaki, M., and M.N. Stavrakakis. 2017. *The Cretan Grapes*, Tropi Publications.

12

The Message on the Bottle

Albert Leonard, Jr.

Admittedly, taste is a very subjective and personal thing. From the very beginning of time, however, it has been recognized that some things were not just different, some were better! And understanding the reasons why that was the case was definitely worth passing on to others to help them distinguish between items that were good and those that were not so good.

The as-yet undeciphered signs on the Minoan Linear A tablet discussed in Chapter 5 (Fig. 5.2) show that each of the three wines associated with ideogram AB 131 were different, but what made them different is unknown. It could have been the wine's taste, the wine's origin, a reference to the sender or receiver, or even scribal variation. But whatever it was, this was important information, and an extra effort was made to call attention to those facts and record them. This task would have been done by an extremely important person, because the job description required at least the rudiments of literacy at a time in history when relatively few people possessed them. This important person (and probably a team of scribes) kept track of whatever goods crossed their paths. Transport and storage vessels that were roughly contemporary with the tablets occasionally reveal the nature of their contents through signs incised or painted on their handles or body by these scribes. Such marks could have been made before the vessel was fired, indicating the jars' origin, the workshop from where they came, or something else (that we do not know) that had been predetermined. Rarely, the marks were added after firing, suggesting the jar had a specific function or use. The decipherment of Mycenaean Linear B gave voice to these mute, Minoan ideograms—in Greek! We learn that a group of Hellenic people of the 13th century B.C. considered some of their wines to be "*me-ri-to-ja*" (μελιτιος, *melitios*) or "honey-sweet," and a scribe felt that it was important to call attention to that fact in their records.

In the eighth century B.C., Homer gave context to these scattered words and phrases, weaving stories that would both entertain and educate his audiences. When describing the luxurious vineyards of King Alcinous in the Odyssey, his audience would have immediately understood why he mentioned laborers busily drying the newly harvested grapes in the sun (Homer, *Odyssey* 7:120–127). They would have known that this extra, time-intensive effort would ensure a sweeter (and hence more desirable) wine. And the containers in which this special wine was stored may have been marked in such a way as to distinguish the "premium" from the more common wines. This was certainly the case in Odysseus's own storeroom. When his son Telemachus was preparing for his trip to visit Menelaus, he went down to the storeroom where "great jars of wine, old and sweet . . . (were) . . . (ar)ranged in order along the walls" (Homer, *Odyssey* 2:340–350). And his aging nurse Eurukleia had no trouble selecting 12 jars of sweet wine that was "the choicest after" the wine that was being saved for the joyous day that Odysseus returned home to Ithaca (Homer, *Odyssey* 3:350–355). Even in this early period, the fame of the vintner was important. When Odysseus spoke of the 25-year-old wine with which he hoped to charm the cyclops Polyphemus, he named and even bragged about the vintner who made his wine. This special drink was made by Maron, the semi-divine son of a Cretan princess, who had served as a priest of Apollo in Ismaros, a famous wine-growing area in Thrace. Knowing details about a wine––how and where it was made and by whom––has always been important to the consumer. Maron's wine was so strong, it had to be cut 20:1 with water before imbibing. It is unfortunate that no one told the cyclops (Fig. 12.1)!

Figure 12.1. Greek terracotta figurine depicting the cyclops Polyphemus relaxing with Odysseus's gift: a cup of 25-year-old wine made in Ismaros (Thrace) by the legendary winemaker Maron. Perhaps from Boeotia, late 5th to early 4th century B.C., height 13.6 cm. Boston, Museum of Fine Arts, acc. 01.7760. Photo J. Johnson, https://en.wikipedia.org/wiki/Polyphemus#/media/File:Polyphemos_reclining_and_holding_a_drinking_bowl.jpg, accessed May 27, 2025; CC BY-SA 4.0, https://creativecommons.org/licenses/by-sa/4.0/.

The sharing of this type of information was also important in Classical times. For example, in Epigrams 106 and 108, Martial compares the quality of different wines (see Ch. 1, p. 11). His audience would probably have been familiar with the reputation of the vineyards around Knossos (Trainor 2021), and others would have heard of the renowned century-old Falernian wine. In fact, many may have enjoyed a cup or two of these wines at the *thermopolium* (neighborhood bar; Fig. 12.2). Whether the famous Monemvasios oinos (Monemvasia or Malvasia wine, see Ch. 10) received its accolades from the region in which it was produced or from the name of the port through which it was traded, consumers knew that this brand stood for quality, and had done so for over four centuries.

Aside from its reputation, information about a wine could also be gleaned from the shape, structural details, and markings of the transport containers in which Greek, Roman, and even Byzantine wines traveled from port to port around the Mediterranean World. The vinicultural rivalry of competing wine-producing areas was sometimes displayed on amphorae, from seals proudly stamped into the soft clay of their handles, before the vessel had been sent to the kiln for firing. In rare cases, these sealings provided the date of the product (its vintage), as well as its place of origin.

Unfortunately, the degree of wine connoisseurship displayed by the Venetians slowly eroded as more and more of their world dissolved into the Ottoman Empire. The centuries-old practice of providing the wine consumer with information about the source of his purchase, and, hence, an idea of the wine's quality, had all but ceased in Crete during the 19th and 20th centuries. Knowledge of the quality of a specific wine no longer traveled far beyond the fame of its region, grape, or winemaker. That is, until an intrepid force arrived upon the Greek wine scene.

Stavroula Kourakou-Dragona learned about oenology (the study of wine and wine making) in France, earned a doctorate in chemistry, and for six decades worked tirelessly with Greek and other European

Figure 12.2. A painted sign at the *Ad Cucumas* (At the Jars) wine shop in Herculaneum, Italy, advertised a selection of four different wines, each with its own price. The town was destroyed by the eruption of Mt. Vesuvius in A.D. 79. Photo C. Raddato, https://commons.wikimedia.org/wiki/File:Wine_selling_advertisement_and_prices,_%22Ad_Cucumas%22_shop,_ancient_roman_painting_in_Herculaneum,_Italy_(45563843984).jpg, accessed May 27, 2025; CC BY-SA 2.0, https://creativecommons.org/licenses/by-sa/2.0/.

governments to produce the original Appellation of Origin system for Greek wines. As with the similar system in France, it allowed Greek winemakers to expand their markets well beyond their immediate neighborhoods. Her early work began restoring the reputation of Greek wines internationally, and she continued to be a major influence in educating the public about the wines of Greece (as well as schooling nascent wine writers such as the present author). The modern Greek appellation system would not be where it is today without her knowledge, skills, and perseverance. So what does it mean when you see a cryptic message (PGI or PDO) on the label of a bottle of modern Cretan wine (Fig. 12.3)?

PGI Crete

It is admitted, from the start, that this can be a confusing subject, especially to those approaching it for the first time, or for those who just want to find a good glass of wine. This brief introduction to modern Cretan appellations below does not try to cover everything. It simply attempts to provide an understanding of the thinking that underlies the system, and how that system is designed to work. Clarification, correction, or expansion of any of these points can be found on the excellent Wines of Crete website (www.winesofcrete.gr) that includes a (free) e-book on the subject written by Yannis Karakasis, MW (Karakasis 2021).

The abbreviation PGI (Protected Geographical Indication) stresses the relationship between a product and its place of production. To paraphrase the World Intellectual Property Organization (www.WIPO.int), a "geographical indication" (GI) identifies a product whose special

Figure 12.3. The front and back labels for Lyrarakis Winery's 2021 *Vidiano* wine, just as they came from the printer, demonstrate how much important and welcome information can appear on a modern Greek wine label. Photo courtesy K. Lyrarakis.

quality, character, and/or reputation is due in great part to its origin, and that name (and hence the relationship) must be correctly stated. Crete is one of eight named vinicultural regions in Greece, and the label "PGI Crete" ensures that the wine was made from specific (named) grapes that were grown on the island, and that their vinification was also conducted within that region as well (i.e., on Crete, not on the Peloponnesus or elsewhere).

Within this framework of PGI Crete there are four districts, organized along the lines of the *territoria* (territories) under which Crete was administered by the Venetians, albeit with some variations in names and spellings (Fig. 12.4). Regulations for districts are a little more specific than those for a region and are more tightly tied to the local geography. From west to east there are four or five districts (depending on how you count them). They are PGI Chania-Kissamos, PGI Rethymnon, PGI Chandaka-Herakleion, and PGI Lasithi-Siteia. Finally, within a given district, there may be PGI areas that are smaller than districts. Their regulations are more sensitive and restrictive, as they attempt to capture the essence of a particular locality and to pass on to the consumer the reason why this information is important. After all, wasn't that what the Mycenaean scribe was doing so many centuries ago?

So, in summary, a PGI Crete wine will display on its label the variety of the grape(s) and any viticultural methods that pertain especially to the region of Crete (Fig. 12.5). A PGI Herakleion appellation will define wines that are particularly specific to the Herakleion district in central Crete, whereas a PGI Peza label will indicate that the wine originates in the area of Peza, which is a subset of the Herakleion district, which itself is part of the region of Crete. Easy!

Figure 12.4. For over four centuries, Venice controlled the island of Crete through the administration of four *territoria* (territories). From west to east they were Chania, Rethymnon, Candia, and Siteia, subdivisions that underlie the modern Cretan wine appellation system, which was first determined by S. Kourakou-Dragona. Map Nicolas Vissher I, 1618–1679; photo D. Jansz van Santen, https://commons.wikimedia.org/wiki/File:Atlas_Van_der_Hagen-KW1049B12_098_2-INSULA_CANDIA_olim_CRETA.jpeg, public domain; accessed May 27, 2025.

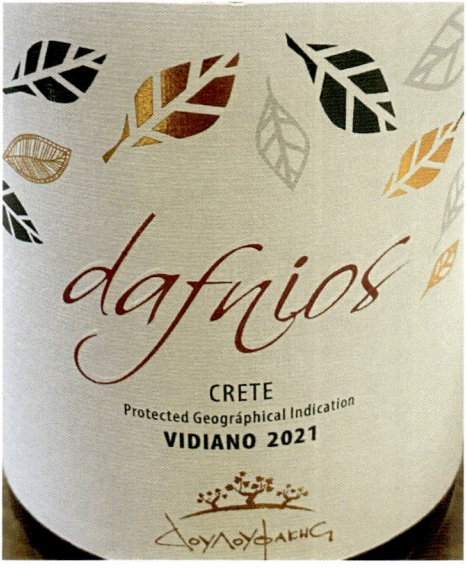

Figure 12.5. The label on Douloufakis Winery's *Dafnios* wine, which uses monovarietal Vidiano grapes, proudly proclaims its PGI (Protected Geographical Indication) status, thereby assuring the customer that Vidiano is a certified varietal grape that is authorized for the island of Crete, and that the grapes were grown and vinified there in the requisite manner. Photo A. Leonard.

PGI Chania

PGI Chania covers all the vineyards over 30 meters above sea level (m.a.s.l.) in the prefecture (administrative district) of Chania. Wine types include the full range allowed, but emphasize the grape varietals listed below with a Cretan legacy (over international grape types). The PGI designation has special regulations concerning the sweetening of a wine by natural sun drying (raisining) or fortification.

> White wine: dry, semi-dry, semi-sparkling dry, semi-sweet, sweet, sweet from dried grapes, and sweet fortified
> Rosé wine: dry, semi-dry, semi-sweet, sweet, and sweet from dried grapes
> Red wine: dry, semi-dry, semi-sweet, sweet, sweet from dried grapes, and sweet fortified
> White varietals: Athiri, Thrapsathiri, Vilana, and White Muscat
> Rosé varietals: as white or red wines
> Red varietals: Fokiano, Kotsifali, Mandilari, Romeiko, and Tsardana
> International varietals (mixed): Cabernet Sauvignon, Carignan, Grenache Rouge, Macabeo, and Syrah

PGI Kissamos

PGI Kissamos is a designation established in 1990 in the district of Chania. It is the home of the local (traditional), high-alcohol Marouvas-style wine. The grape varietals listed below emphasize Cretan-legacy (over international) grapes, and the PGI designation notes that percentages of Romeiko may be limited.

> White wine: dry
> Rosé wine: dry
> Red wine: dry
> White varietals: Athiri, Thrapsathiri, and Vilana
> Rosé varietals: Athiri, Grenache, Romeiko, Thrapsathiri, Ugni Blanc, and Vilana
> Red varietals: Romeiko
> International varietals: Cabernet Sauvignon, Carignan, Grenache, and Syrah

PGI Herakleion

PGI Herakleion covers all the vineyards over 30 m.a.s.l. in the prefecture (administrative district) of Herakleion. Established in 1989, it is the largest of the four vinicultural regions of Crete and contains three contiguous PDO (Protected Designation of Origin, see below) appellations, from west to east: PDO Dafnes, PDO Archanes, and PDO Peza.

> White wine: dry, semi-dry, semi-sweet, sweet from dried (raisined) grapes
> Rosé wine: dry, semi-dry, semi-sweet
> Red wine: as white wines
> White varietals: Athiri, Daphni, Malvasia di Candia Aromatica, Plyto, Thrapsathiri, Vilana, Vidiano, and White Muscat
> Rosé varietals: Kotsifali, Ladikino, Liatiko, Mandilari, and White Muscat.
> Red varietals: Kotsifali, Ladikino, Liatiko, and Mandilari
> International varietals (mixed): many, including Cabernet Sauvignon, Chardonnay, Sauvignon Blanc, and Syrah.

PGI Rethymnon

PGI Rethymnon is a designation established in 2010 that covers almost all the vineyards over 30 m.a.s.l. in the prefecture (administrative district) of Rethymnon. Grape varietals listed below emphasize Cretan legacy (over international) grapes. Note that special regulations exist concerning the sweetening of a wine by natural sun drying (raisining) or fortification.

> Dry white wine: semi-dry, semi-sweet, sweet
> Dry rosé wine: as white wines
> Dry red wine: as white wines
> White varietals: Athiri, Vidiano, Vilana, Thrapsathiri, and White Muscat
> Rosé varietals: as white or red
> Red varietals: Kotsifali, Liatiko, Mandilari, and Romeiko
> International varietals: Cabernet Sauvignon, Grenache Rouge, and Syrah

PGI Lasithi-Siteia

PGI Lasithi-Siteia was established in 1989. This designation pertains to all the vineyards over 30 m.a.s.l. in the prefecture (administrative district) of Lasithi, including PDO Sitia Red and PDO Sitia White wines.

> White wine: dry, semi-dry, semi-sweet
> Rosé wine: dry, semi-dry, semi-sweet
> Red wine: dry, semi-dry, semi-sweet
> White varietal: Assyrtiko, Athiri, Plyto, Thrapsathiri, Vilana, and White Muscat
> Rosé varietal: as white or red wines
> Red varietal: Kotsifali, Ladikino, Liatiko, and Mandilari
> International varietals: Cabernet Sauvignon, Carignan, Grenache Rouge, Merlot, Syrah, Trebbiano, and Viognier

PDO Crete

The appellation PDO (Protected Designation of Origin) adds the further stipulation that a product has been made according to recognized (usually historic or traditional) methods practiced in that geographic area. These terms may be confusing, and even the differences in qualification between a PGI and a PDO wine may appear to be contradictory at times. But remember that these rules are designed to help the producers accurately define their product and to enhance the consumer's experience (Fig. 12.6). They are intended to protect the integrity of the product, the producer, and the consumer, and are briefly listed below, alphabetically.

> PDO Archanes, established in 1971, is the central PDO in the Herakleion district and is restricted to dry red blends of Kotsifali and Mandilari(a).
> PDO Daphnes, is the westernmost PDO in the Herakleion district and is applied to dry or sweet wines made from the Liatiko grape.
> PDO Chandakas-Candia, established in 2011, covers most of the Herakleion district. The appellation includes dry red blends of Kotsifali (70%) and Mandilari (30%), as well as dry white wines, of which 85% must be Vilana with the remaining 15% consisting of Assyrtiko, Athiri, or Thrapsathiri.
> PDO Malvasia Chandakas-Candia, established in 2011, covers approximately the same area as PDO Chandakas-Candia, but also authorizes white sweet (sun-dried and sun-dried-fortified) wines to be made from a blend of a minimum of 85% Assyrtiko, Athiri, Thrapsathiri, and/or Liatiko with the remaining 15% consisting of White Muscat or Malvasia di Candia Aromatica. It is required that these wines be aged in oak for a minimum of 24 months.

Figure 12.6. Domaine Oikonomou released this 2015 *Sitia* wine as a PDO (Protected Designation of Origin) wine, certifying that it was made from the authorized blend of Vilana to Thrapsathiri grapes, that they were grown according to the regulations of the Siteia prefecture (East Crete), and that vinification had been conducted in the traditional manner of that area. The label also carries the older terminology *Onomasia* (Appellation) *Proelefsis* (Origin) *Anoteras* (Superior) *Poiotitos* (Quality), together often abbreviated as OPAP, as well as the abbreviation *V.Q.P.R.D.* (Vin de Qualité Produit dans une Région Déterminée), to reinforce the fact that this is a quality wine produced in a determined (specified) region. Photo A. Leonard.

PDO Peza, is the easternmost PDO in the Herakleion district. It was established in 1971 for dry red blends of Kotsifali and Mandilari(a) and in 1982 for dry monovarietal Vilana white wines. There are elevation restrictions on the vineyards of Vilana.

PDO Siteia-Lasithi, established for red wines in 1971, includes dry red blends of small-berried Liatiko (minimum 80%) and Mandilari (20%), as well as sweet monovarietal Liatiko wines with specific stipulations on fortification. In 1989 this PDO was extended to include dry white wines made from blending Vilana (80%) and Thrapsathiri (20%) grapes grown on the upland plateau.

PDO Malvasia Siteia-Lasithi, established in 2011, covers many of the high altitudes in the district and includes sweet white wines made from sun dried grapes (and fortified-from-the-sun dried grapes) that are made with a minimum of 85% Assyrtiko, Athiri, Thrapsathiri, and Liatiko, and a maximum of 15% White Muscat or Malvasia di Candia Aromatica. These wines must spend at least 24 months in oak.

Further Reading

Dalby, A. 2003. *Flavours of Byzantium*, Prospect Books.

Karakasis, Y. 2021. *The Wines of Crete: A Terroir Report*, Herakleion Chamber of Commerce & Industry, https://karakasis.mw/%cf%84he-wines-of-crete-terroir-report/, accessed May 13, 2025.

Kourakou-Dragona, S. 2015. *Vine and Wine in the Ancient Greek World*, trans. M. Relaki, Foinikas Publications.

Lazarakis, K. 2018. *The Wines of Greece*, Infinite Ideas Limited.

Manessis, N. 2000. *The Illustrated Greek Wine Book*, Olive Press Publications.

Murray, A.T., trans. 1919. *Homer: Odyssey*, vol. I: books 1–12 and vol. II: books 13–24 (*Loeb Classical Library* 104 and 105), G.E. Dimock, rev., Harvard University Press.

———, trans. 1924. *Homer: Iliad*, vol. I: books 1–12 (*Loeb Classical Library* 170), W.F. Wyatt, rev., Harvard University Press.

———, trans. 1925. *Homer: Iliad*, vol. II: books 13–24 (*Loeb Classical Library* 171), W.F. Wyatt, rev., Harvard University Press.

Stavrakaki, M., and M.N. Stavrakakis. 2017. *The Cretan Grapes*, Tropi Publications.

Trainor, C.P. 2021. "The Late Hellenistic Wine Press Excavations from Knossos: The Early Iron Age, Hellenistic and Early Roman Contexts," *Annual of the British School at Athens* 116, pp. 235–290.

13

Glimpses of the Past in the Future of Cretan Wines
Resins, Raisins, Pots, and People

Albert Leonard, Jr.

The grape, like any other fruit, is primarily a food for human and animal consumption. But for humans it has a secondary, hidden benefit: it can be a psychotropic stimulus (alcohol) that many people today consider to be a valuable social lubricant. Perhaps it was noticed first while watching the reaction of feeding birds, but once this feature had been observed, recognized, and enjoyed by our early ancestors, they devoted a considerable amount of time and thought, first to enhancing the product itself, and then to extending its availability throughout the year. When not done right and grapes are exposed to oxygen, wine will quickly sour and, at best, turn to vinegar (*vin aigre,* sour wine). That may have been a useful condiment, but it lacked the social benefits and rewards for which so much time and energy had been invested. Let us then, in conclusion, discuss what Cretan winemakers have done right for millennia, and turn our gaze to the future of Cretan winemaking, with an eye to the way that several of the island's traditional grapes, and many of the vinicultural practices employed in the past, are being understood, revived, and improved by the winemakers of today (Figs. 13.1, 13.2).

Resins

Modern archaeological science has demonstrated that the addition of resin (*Pistacia terebinthus, Pistacia atlantica,* etc.) to wine as a preservative (or as a coating to the clay vase to keep the porous container from leaking) has been practiced widely as early as the Stone Age (prior to 3000 B.C.; see McGovern 2003). This was the case at the Early Minoan IA site of Aphrodite's Kephali where a small hilltop fort, overlooking the modern village of Episkopi, guarded the border between southern and northern East Crete (Fig. 1.2:3). Analysis of a large pithos from this site has shown that the wine it held was flavored with resin (see Ch. 3). This use of resins continued throughout the Bronze Age across the eastern Mediterranean. Winemaking communities that lacked local access to the variety of resins so prevalent along the Levantine coast were eager to import them. And there was a ready market to fill that need. Approximately one ton of *P. terebinthus* (Fig. 13.3) was found packed in Canaanite jars (amphorae) on the late 14th-century B.C. ship that went down in a storm at Uluburun, just off the southern coast of Anatolia (Fig. 1.7). Granted, resinous materials served other functions, but the probability that a portion of such a tremendous quantity was intended for the manufacture of wine cannot be excluded. Pliny (A.D. 23–79) tells us that similar resinous materials were added directly to the fermenting wine and that they also were

Figure 13.1. *Kretikos* wine label from Boutari Winery, advertising the PGI (Protected Geographical Indication) of the wine. For seven centuries or more the Vilana grape has been used to make wine in and around the prefecture of Herakleion, but the winemakers of the area are constantly at work to perfect it. Photo courtesy Boutari Winery, Crete. For the image of the Ladies in Blue, see Fig. 11.19.

Figure 13.2. *Minos Palace* wine label. Pride in the island's Minoan past—as illustrated by an adaptation of the Procession Fresco from the neighboring Palace of Minos at Knossos—has been featured on at least one Minos wine label a year since the practice was begun in the 1950s. Photo courtesy N. Miliarakis. For the Procession Fresco, see Fig. 11.1.

Figure 13.3. The antibacterial properties of *Pistacia terebinthus*, also known as the turpentine tree, appear to have been well known to the people of the ancient Mediterranean world. The plant was also recognized for its antioxidative value in extending the shelf life of wines. Photo J. Martin, https://commons.wikimedia.org/wiki/Pistacia_terebinthus#/media/File:Pistacia_terebinthus_SierraMadrona1.jpg, accessed August 1, 2025; CC BY-SA 4.0, https://creativecommons.org/licenses/by-sa/4.0/.

used to seal the containers (amphorae) in which the wine was stored or shipped (Pliny, *Natural History* 14.25.12). To that end, he presented a long, prioritized list of the specific resins that he considered best, with a commentary on their individual benefits (Pliny, *Natural History* 14.24.1–9). By keeping out the oxygen, it was hoped that the wine would remain palatable longer.

As the Romans advanced into northern Europe and encountered cultures transporting liquids in wooden barrels, the heavy use of resins lessened in the western Roman empire, where more personal-sized, barrel-shaped glass jugs also enjoyed considerable popularity (Figs. 13.4, 13.5). This was not the case, however, in the east, where the old traditions carried on. This contrast between eastern and western palates was sharply drawn in the 10th century by Liutprand of Cremona in describing the wines that he was forced to endure during a visit to Byzantine Constantinople. They were simply undrinkable! (see Dalby 2003).

The topic of resinated wines reared its head again after World War II, when tourists began to visit Greece. Here they found that most of the wine available in Attica was made from the juice of Savatiano (Fig. 13.6) and/or Roditis grapes that had been treated to varying percentages of mastic or pine pitch. Often, this "retsina" had been fermented in the very barrel from which it would be drawn in the basement of a restaurant. By the 1960s retsina had begun to be bottled in glass, the most distinctive brand being offered by Kourtaki Winery with its iconic yellow label and crown cap. By the time that tourists ventured out to enjoy the beauty that Greece had to offer, they found that retsina—now a term used as a substitute for any Greek wine—had preceded them, stigmatized by references to it as tasting of "turpentine" and "paint remover." This sad situation was followed in the 1970s by the scourge of phylloxera (an insect that attacks grapes), forcing many vineyards to uproot their traditional grapes and replace them with international varietals that were ill-suited to the Cretan *terroir*.

Things started to change for the better in the late 1990s when Yannis Paraskevopoulos of Gaia Winery (at Nemea in eastern Peloponnese) released a much more delicate (Roditis-based) retsina named *Ritinitis*

Figure 13.4. While the amphora was the preferred method by which to ship wine around the Mediterranean, the wooden barrel was the preferred means on the rivers of Europe, as can be seen in this reproduction of a sandstone funerary monument excavated at Neumagen-Dhron, Germany. The original sculpture memorialized a Roman wine-trader, ca. A.D. 220. Rheinisches Landesmuseum Trier, Germany. Photo C. Raddato, https://commons.wikimedia.org/wiki/File:Funerary_stone_monument_found_in_Neumagen_in_the_shape_of_a_rowing_ship_for_transport_of_wine_barrels_on_the_Moselle_river,_about_220_AD,_Rheinisches_Landesmuseum_Trier,_Germany_(29411833330).jpg#, accessed May 27, 2025; CC BY-SA 2.0, https://creativecommons.org/licenses/by-sa/2.0/.

Figure 13.5. Glass jugs and juglets in the form of barrels were extremely popular in northern and western Europe, especially from the second through the fourth centuries A.D. They are commonly called "Frontinus bottles" after the name (or its abbreviation) that often appears on their mold-blown bases. Height 11.6 cm. New York, The Metropolitan Museum of Art, gift of Henry G. Marquand, 1881, acc. 81.10.73, https://www.metmuseum.org/art/collection/search/245230, accessed August 1, 2025; CC0 1.0, https://creativecommons.org/publicdomain/zero/1.0/.

Nobilis and demonstrated the tremendous potential of Greek wines. About the same time, he produced *Thalassitis* (Seaborn), a monovarietal wine from the Assyrtiko grape, a varietal that has strong historical connections to Santorini (Thera), Crete, and the Aegean Islands (Fig. 1.6).

Today on Crete, a dry white retsina is released under the Traditional Appellation designation. Michalakis Estate makes two versions: *Lato Retsina*, in a 750 ml bottle with a cork closure, and *Fina Retsina*, in a 500 ml bottle with a crown cap. Both wines are flavored with natural pine resin and are low in alcohol (11% abv.). Other offerings include Pnevmatikakis Estate (in Kissamos) that crafts their *Retsina* wine from Vilana grapes with pine resin sourced from a natural forest in northern Euboea that is added during fermentation. The wine is released in a 500 ml bottle with a screw cap (11.5% abv.). Titakis Winery in Peza produces their *Retsina* from 100% Rozaki must that is fermented in traditional, open concrete tanks and released in squat 500 ml glass bottles (11.5% abv.). In the past, the Rozaki/Razaki grape (a synonym of the Turkish Çavuš grape) was planted more widely as a table grape than as a wine grape (see Stavrakaki and Stavrakakis 2017; Robinson, Harding, and Vouillamoz 2012). The grape's origin is complicated, but it definitely comes from the east. Note that two of the other synonyms for the Çavuš grape—Regina di Beyrouth (in Italy) and Dattier de Beyrouth (in France)—both contain the name of the capital port city of Lebanon, homeland of the Phoenicians; and that, in Romania, the grape is known as Aleppo, the homeland of the Aleppo pine (*Pinus halapensis*), a popular choice for flavoring wines in antiquity. On Crete today, retsina can boast a lengthy vinicultural pedigree. Given the fact that soft drink options include *Mast*, self-described as "a refreshing, sparkling drink with mastic," suggests that the future may be bright for retsina's return to popularity on the Great Island.

Figure 13.6. A bunch of Savatiano grapes. Photo courtesy M. Stavrakaki and M.N. Stavrakakis.

Raisins

In recent years, more and more winemakers have come to believe that, at least figuratively, a great wine is born in the vineyard long before it reaches the winery, and it appears that the more observant winemakers of the past felt the same way. The earliest literary description of a method to "extend the shelf life" of a wine while it was still in the vineyard was by Homer in the eight century B.C. when describing the fertile gardens of the Phaeacians, who had provided Odysseus refuge after he escaped from Calypso's Island. "There, too, is his fruitful vineyard planted, one part of which, a warm spot on level ground, is being dried in the sun, while other grapes men are gathering, and others, too, they are treading; but in front are unripe grapes that are shedding the blossom, and others that are turning purple (Homer, *Odyssey* 7.123–126).

Slightly later, more of the details of this process were included in the personal instructions that Hesiod gives to his brother Perses (Fig. 13.7). "When Orion and Sirius are come into midheaven, and rosy-fingered Dawn sees Arcturus, then cut off all the grape-clusters, Perses, and bring them home. Show them to the sun ten days and ten nights: then cover them over for five, and on the sixth day draw off into vessels the gifts of joyful Dionysus" (Hesiod, *Works and Days* 609–614).

Here Hesiod links the time of the grape harvest to astral events that occur late in the growing season, when the sugar content of the grapes would be approaching its highest level. Combining a late harvest with the practice of allowing the grapes to sun dry on mats or racks above ground would further concentrate the sugars and result in a sweeter wine. Covering the grapes for a portion of that time could help to protect the grape clusters from mold (Fig. 13.8).

Sweet wines, also referred to as dessert wines, are often said to be crafted in the Phoenician (or Punic) manner because many details of the process were given by a writer in ancient Carthage (modern Tunis), a colony traditionally founded by the Phoenicians in 814 B.C. His name was Mago, and ca. 500 B.C. he wrote a multi-volume treatise on agriculture in the Phoenician/Punic language in which he gives his recipe for making the best *passum* (a past perfect participle of the Latin verb *pandere*, "to spread out to dry"), the name by which this type of wine has been known for centuries.

His work served as the "go-to" textbook for winemaking throughout antiquity. In fact, Mago's writings had so much valuable information on viniculture that the victorious Romans, after they set Carthage ablaze at the end of the Punic Wars (146 B.C.), rescued every copy of his manuscript that they could find and, by the order (and at the expense) of the Roman Senate, they were translated from the original Phoenician or Punic language into Latin (and later into Greek). Unfortunately,

Figure 13.7. A mosaic "portrait" commemorating Hesiod (*Esio-dus*), the author of *Works and Days*, was excavated at a third century A.D. Roman *domus* (house) in the city of Augusta Treverorum (modern Trier), Germany. Rheinisches Landesmuseum Trier, the Monnus Mosaic. Photo C. Raddato, https://commons.wikimedia.org/wiki/File:Monnus_Mosaic,_detail_of_Hesiod_(ESIO-DVS)_from_a_Roman_Domus_in_Augusta_Treverorum_(Trier),_end_of_the_3rd_century_AD,_Rheinisches_Landesmuseum_Trier,_Germany_(30038562895).jpg, accessed May 27, 2025; CC BY-SA 2.0, https://creativecommons.org/licenses/by-sa/2.0/.

> **Late Harvest Wine vs. Raisined Wine**
>
> The term "late harvest" refers to wines made from grapes that have remained attached to the vine until late in the season, providing that the flow of nutrients to them has not been interrupted by manually twisting (*torculare*) or other means. This late harvest will allow the sugars to concentrate and make the wine easier to preserve, but, early on, this technique can result in a loss of acidity as the grapes begin to dehydrate. In "raisined" wine, as the term is used here, the grapes are dried "off the vine," manually separated, cut off from any nourishment that they have been receiving through the vine. The grape clusters are spread out (in Latin, *pando*) on loosely woven, straw mats that allow air to circulate in and around the clusters. This practice will completely stop the ripening process, and *both* the sugars and the acidity will be intensified, hopefully producing a well-balanced wine. Many wineries now use straw-lined plastic or stacked wooden racks to provide this ventilation. The greatly reduced yields from hand sorting the fruit, the labor-intensive process, and the fairly large percentage of liquid lost through evaporation combine to make these wines very expensive in the modern market. We can assume that the same held true on the docks of Rethymnon, the taverns of Monemvasia, and along the quays of La Serenissima (Venice).

Figure 13.8. Another detail of the mosaic from the same Roman house in Trier (see Fig. 13.7) depicts Octob(er), the eighth month of the Roman calendar, as Dionysos/Bacchus wearing an ivy wreath on his head and carrying the *thyrsus* (magical fennel staff) by which he is identified. Rheinisches Landesmuseum Trier, Germany, the Monnus Mosaic. Photo C. Raddato, https://commons.wikimedia.org/wiki/File:Monnus_Mosaic,_square_panel_detail_with_personification_of_the_month_of_October_(Bacchus_with_thyrsus_and_wreath),_from_a_Roman_Domus_in_Augusta_Treverorum_(Trier)_(29924872902).jpg, accessed May 27, 2025; CC BY-SA 2.0, https://creativecommons.org/licenses/by-sa/2.0/.

neither a complete copy of Mago's original work, nor its translations, have survived, but major sections of it were preserved by Columella, a first century Roman writer, in an extensive work on farming entitled *De re rustica* (Fig. 13.9). A section of that work is instructive here:

> Mago gives the following instructions for excellent passum, and I [Columella speaking] have made it this way myself. Harvest well-ripened very early bunches of grapes; reject any mildewed or damaged grapes. Fix in the ground forked branches or stakes not over four feet apart, linking

them with poles. Lay reeds across them and spread the grapes on these in the sun, covering them at night to keep the dew off. When they have dried, pick the grapes, put them in a fermenting vat or jar and add the best possible must so that they are just covered. When the grapes have absorbed it all and have swelled, after six days, put them in a basket, press them and collect the passum. Then tread the pressed grapes, adding very fresh must made from other grapes that have been sun-dried for three days. Mix all this and put the mixed mass through the press. Put this *passum secundarium* into sealed vessels immediately so that it will not become too harsh (*austerum*). After twenty or thirty days, when fermentation has ceased, rack (the wine) into other vessels, seal the lids with gypsum and cover them with skins. (Columella, *De re rustica* 12.39.1).

Despite this attention to detail, neither the early Cretan farmers, nor their Phoenician visitors, could have correctly understood the scientific complexities of microbial fermentation. In the days before Louis Pasteur (1822–1895), they would not have known that unseen (microscopic) yeasts from the skins of their grapes would have fed on the sugars in the grape juice to produce the desired alcohol. But both Homer and Hesiod seemed to sense that increasing the sugar content by raisining the grapes would not only make the wine more palatable (sweeter), but also would make the wine last longer, enabling it to be traded farther. This desirable combination of age and sweetness is exactly how Homer describes the special wine that Telemachus selects from his storeroom to bring on his voyage to sandy Pylos. "There . . . stood great jars of wine, old and sweet, holding within them an unmixed, divine drink" (Homer, *Odyssey* 2.340). Each attribute is a factor of the other. Sweet (i.e., raisined) wines last longer and they travel better. And when you are ready to serve them, mixing the wine with water will bring the alcohol level down while still retaining the sweetness. Wine was important to the manners and rituals of wine-consuming cultures of the elite Mediterranean World, of which the Levant had long been an integral part. Homer's audience would have completely understood this reference to the wine, its context, and possibly its eastern associations. At the funeral games for Patroklos (friend of Achilles who died in the Trojan War), one of the prizes was a silver krater (mixing bowl) described as "a masterpiece of Sidonian [i.e., Phoenician] craftmanship" (Homer, *Iliad* 23.741–744), and, as a parting gift from his fact-finding visit to Sparta, Menelaus had presented young Telemachus with an item that he

Figure 13.9. A statue honoring Columella stands in the Plaza de las Flores (Square of the Flowers) in Cadiz, Spain. He overlooks the flowers and other agricultural products about which he wrote. Photo A. Leonard.

claimed was the costliest in his storeroom: a silver and gold krater that had been crafted by Hephaistos for Phaedimos, King of Sidon (Homer, *Odyssey* 4.609–619). You might carry the wine to the party in a rustic wineskin, but proper service required mixing it with water in a krater before it was consumed (Homer, *Odyssey* 9.196).

A good case can be made that the technique of raisining grapes, and perhaps some of the grape varietals themselves, originated in places east of Crete and later "island-hopped" westward in the holds of Phoenician or mixed-registry ships. Did transient Phoenician winemakers share their vinicultural secrets with their Cretan hosts, as they certainly shared their alphabet?

Today, wines produced in a similar fashion continue to be made in places that enjoy a warm climate. In France they are called straw wines (*vin de paille,* or wine of the hay), in Italy *Vino Santo* (holy wine), and in Greece, *Vinsanto* (holy wine; also sometimes *Vissanto*), to name a few. Their common denominator is the fact that the grapes had first been dried on straw mats or rooftops to increase the level of sugar and make the wines last.

On Crete, another term is used for raisined wine: *liastos*, meaning sun dried, with reference to the work of Helios, the sun or sun god. This wine is a specialty of the island of Tinos (Fig. 1.6), which from 1207 until 1715 had formed part of the Kingdom of Candia/Crete. On the island of Cyprus to the east (Fig. 1.7)—the commercial link between the Levant and the Aegean world—an equivalent wine was known as *Nama*, and its fame also rested in part on its sweetness. It was literally fit for a king. This was the wine with which King Richard I of England (*Coeur de lion* or Richard the Lionheart) celebrated his wedding to the comely Berengaria of Navarre. It had been made on the island by his knights at *La Grande Commanderie* (their military quarters and estate located close to Limassol in eastern Cyprus). Because of its excellence, the wine quickly became known as Commandaria, a title that is still used today. Some would say that this wine was the first to have earned its own *appellation d'origine contrôlée* (AOC, or controlled designation of origin).

In their *Liastos*, Lyrarakis Winery blends six of the historical grapes of Crete (listed here alphabetically): 30% Assyrtiko, 10% Dafni, 10% Plyto, 15% Thrapsathiri, 26% Vidiano, and 9% Vilana, all of which could have lived together comfortably as a field blend in some ancient Cretan vineyard. These grapes are dried "in the shadow" for seven days and in direct sunlight for two more. Following fermentation, the wine is aged in 225-liter oak barrels for 12 months. Lyrarakis' *Liastos 2015* offering was released as a PGI Crete wine (11% abv. with 164 g/L of R.S. and 7.2 g/L of acidity). Silva-Daskalakis Winery's organic, naturally grown *Liastos* is a monovarietal Liatikos wine made from grapes that are dried in the sun for six to eight days before being fermented on local yeasts, after which it rests in French and American oak barrels for 15 months before being transferred to 500 ml bottles (13% abv). Both are excellent examples of wines that have benefited from 5,000 years of history.

Yeasts

In describing a wine's *terroir*, we are talking about the many local elements that combine to make that wine different and special. These elements might include topography, microclimate, soil type, water availability, sun exposure, and diurnal temperatures. Rarely mentioned is the most local element after the grape itself: the tiny (microscopic) yeast cells that are at home in that vineyard. They are responsible for starting the process of fermentation and, hopefully, capable of moving things along to fruition.

Living yeasts occur naturally on the skins of fruit, including the grape. When that skin is pierced, whether by a bird's beak, human feet, or just the weight of fellow grapes in a container, these yeasts are introduced to the sugary juice of the fruit. At that point, fermentation begins as the yeasts feed on the sugar and convert it to alcohol (ethanol) and carbon dioxide (CO_2).

There are many different strains of yeast, each of which will contribute something (a plus or a minus) to the final product. The best and most important yeast for winemaking (and baking) is *Saccharomyces cerevisiae* because of its predictability and its voracious appetite. In comparison, other yeasts are usually grouped together as "non-*Saccharomyces* yeasts" and are often referred to as "wild yeasts." They are unpredictable, frail, and many will die off as the alcohol level approaches 5%. It takes *S. cerevisiae* to come to the rescue and finish the job of building the alcohol level up to 12–14%. A few wild yeasts are stronger and will produce a wine with a higher percentage of alcohol, especially when present in the juice of grapes with high sugar content. This is the case on the island of Ikaria, long associated with the Pramnian wine that Hecamede used to prepare a restorative (κυκεών [*kykeon*], or a drink made from wine or water, barley, and other ingredients) for Nestor and his friends in their brief respite from the war at Troy (Homer, *Iliad* 11:635–641; Fig. 13.10). Afianes Winery of Ikaria consistently produces a natural blend of 85% Fokiano and 15% Kountouro/Mandilari grapes that is released at or above 15% abv. This blend was probably the case with many of the more successful wines in antiquity. Tim Bell, winemaker at Dry Creek Vineyard in Healdsburg, CA, points out (personal communication) that yeasts could also remain present both in the fermentation vessels used and in the general winery environment. So, over time, a particular winemaking location might develop its own unique, and very distinctive, yeast population within the vessels and the buildings themselves.

Figure 13.10. A scene on the interior of an Attic red-figure kylix (drinking cup) from Vulci by the Brygos Painter. Also known as the Iliupersis Cup, it depicts what many interpret as Hecamede pouring a refreshment for Nestor as described by Homer. Ca. 490–480 B.C. Paris, Louvre Museum, acc. MNB 3047; G 152. Photo Bibi Saint-Pol, https://commons.wikimedia.org/wiki/File:Briseis_Phoinix_Louvre_G152.jpg#, public domain, accessed May 27, 2025.

Pots

Winemaking practices that can trace their origins back into distant antiquity are beginning to come together in the current movement toward "natural" or "orange" wines. The modern wine-consuming psyche seems to be drawn to the millennia-old tradition of a long fermentation period in buried ceramic pots, such as those currently enjoying an exciting revival in the Republic of Georgia (Fig. 1.7; see Lordkipanidze, ed., 2017). There, egg-shaped (i.e., without handle) vessels known as qvevri (KVEV-ri), or karas in bordering Armenia, can produce up to 1,000 liters of wine with a minimal amount of attention (Fig. 13.11). Unfortunately, the word qvevri is often (mistakenly) translated as amphora. This can be problematic because amphora is a Greek word (and to quote Gus Portokalos from the film My Big Fat Greek Wedding [2002], "there you go") that can be traced back to the *a-pi-po-re-we* encountered in Bronze Age Linear B (Greek). It has always referred to a two-handled jar, especially one that has been designed purposefully to be *carried* by those two handles.

On Crete, today, the preferred term for an equivalent large ceramic vessel used to ferment grapes seems to be pithari(a), and similar jar(s) are known on the island of Cyprus. Pithari seems to be more popular in the Aegean area, as a whole, but the less-specific term pithos(oi) is also found. The latter word is frequently used in archaeology to describe the large, multi-functional terracotta jar(s) that are ubiquitous at many Minoan (and later) sites on the island (see Ch. 8).

And, while scholars debate the appropriateness or correctness of using the term "amphora" in the creation of ancient wines, Cretan winemakers have gone ahead and crafted their own vessels. Silva-Daskalakis Winery's natural, organic *Grifos* series aims at following the Minoan manner of vinification by fermenting on yeasts (variously described as indigenous, local, and native) in 300-liter clay pitharia as pictured on their label (Fig. 13.12). Their dry white wine is 100% Vidiano (13.5% abv.), their dry rosé is made from 100%, sun-dried Kotsifali (14.5% abv.), and their dry red is a monovarietal Liatiko (15% abv.). Each wine enjoys an oak program specific to the individual grape before it is bottled, unfiltered and without sulfites, in 500 ml glass bottles with recyclable sugarcane stoppers.

Douloufakis Winery, working with an eye toward the island's historic past, also employs pithoi in their wine-making process in its recently inaugurated *Amphora* series, calling to mind the ancient Greek transport vessel. They produce monovarietal Vidiano and Muscat White of Spina wines that are fermented in 250–300-liter pithoi without the addition of yeast, use only a minimum of sulfur, and are matured in stainless steel. The 2020 *Vidiano Amphora* was released at 12.6% abv. with 2.5 g/L of R.S. Their 2020 *Muscat Amphora White* (Fig. 13.13), described as a natural orange wine, was released at 14.1% abv. with 3.4 g/L of R.S. (reduced from the 15.5% abv. of their 2017 vintage).

Figure 13.11. A late 19th-century winemaker leans against a qvevri (large storage jar) in Kakheti, Republic of Georgia. These large, handleless vessels would have been buried in the earth up to their rim and would serve as both fermenters and storage vessels. Photo M. Barry, 1881, https://commons.wikimedia.org/wiki/Category:Kvevri#/media/File:Barry_(capitaine)._F._25._Grand_vase,_pour_la_conservation_du_vin_en_Kacheti_Géorgie._Mission_scientifique_de_Mr_Ernest_Chantre._1881.jpg, public domain, accessed May 27, 2025.

Figure 13.12. The Silva-Daskalaki Winery uses indigenous yeasts to ferment the monovarietal wines of their organic Grifos series in large (300-liter) terracotta jars, as shown on their Liatiko wine label. Image courtesy Silva-Daskalaki Winery.

People

As discussed above, we have seen that the creation of a fine wine is never accidental. It is the result of innumerable, tiny steps that, for our purposes, can be reduced to four: the growth and harvesting of the grapes, the crushing and pressing of the juice (must), the process of fermentation itself, and the storage of the finished product.

The simple paradigm of wine-making begins with the planting of an appropriate grape for the specific *terroir*, the most beneficial way to nourish the vine and manage its canopy, and the decision on when to pick the fruit. After the grapes have been sorted and any damaged fruit is removed, comes crushing (by foot, mechanical, or even gravitational means) to rupture the skins and bring the yeast into contact with the sweet juice of the grapes. Done correctly, this should bring about a change. But the change will not be instantaneous. The ancient winemaker would have spent hours—even days—waiting, staring hopefully at the placid surface of the liquid in his pithos/pithari. Slowly, hesitatingly, the first few bubbles would appear. And then, the once-still grape juice would begin to ferment (from *fervere*, to boil, in Latin) and would continue to bubble violently for several days. Before the discoveries of Pasteur, this action was probably attributed to the presence (or intervention) of the god Dionysos. And when this deity had finished his work and left the scene, it was time for the vintners to complete the job. Speed was of the essence! The finished wine would be racked (moved) from the fermenter into smaller containers, a process that may or may not have included filtering (though a loosely woven cloth) or fining (the addition of egg white to soften the wine by removing some of the tannins, and possibly increasing the clarity of the wine). Finally, it would be "bottled," perhaps in a resin-prepared amphora and sealed with a vegetal stopper. This last phase had to be accomplished quickly because, in the presence

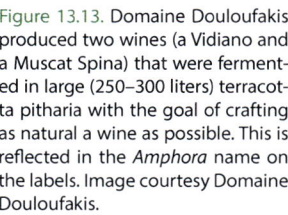

Figure 13.13. Domaine Douloufakis produced two wines (a Vidiano and a Muscat Spina) that were fermented in large (250–300 liters) terracotta pitharia with the goal of crafting as natural a wine as possible. This is reflected in the *Amphora* name on the labels. Image courtesy Domaine Douloufakis.

of oxygen, Acetobacter bacteria could convert the ethanol in the wine to acetic acid (vinegar). After that, the wine's fate would again be in the hands of the gods, until it was tasted.

The crucial elements that are not sufficiently stressed in the paradigm above are the people in the process. Some would have brought with them their technical skills, others contributed their artistic talents, many simply provided their fortitude and tenacity. Through the centuries, the strength, ingenuity, and perseverance of the Cretan people made it work. It has truly been a pleasure to offer a few glimpses of their 5,000 years of failures, struggles, successes, and eventual triumphs. To them, this story of their labor is offered as a tribute.

Further Reading

Afianes, N. 2004. *The Vine Which Is in Ikaria: On the Wine of Ikaria from Antiquity to the Present Day*, Ikariakon Meleton Company, University of the Aegean, pp. 1–74.

Bostock, J., and H.T. Riley, trans. 1855. *The Natural History of Pliny*, Henry G. Bohn.

Dalby, A. 2003. *Flavours of Byzantium*, Prospect Books.

Evelyn-White, H.G., trans. 1914. *Hesiod, Homeric Hymns, Epic Cycle, Homerica* (*Loeb Classical Library* 57), Harvard University Press.

Forster, E.S., and E.H. Heffner, trans. 1955. *Columella: On Agriculture*, vol. III: books 10–12: *On Trees* (*Loeb Classical Library* 408), Harvard University Press.

Goode, J. 2005. *The Science of Wine: From Vine to Glass*, University of California Press.

Johnson, H. 1989. *Vintage: The Story of Wine*, Simon and Schuster.

Lordkipanidze, D., ed. 2017. *Georgia, the Cradle of Viticulture*, Georgian National Museum.

McGovern, P.E. 2003. *Ancient Wine: The Search for the Origins of Viticulture*, Princeton University Press.

Murray, A.T., trans. 1919. *Homer: Odyssey*, vol. I: books 1–12 and vol. II: books 13–24 (*Loeb Classical Library* 104 and 105), G.E. Dimock, rev., Harvard University Press.

———. 1924. *Homer: Iliad*, vol. I: books 1–12 (*Loeb Classical Library* 170), W.F. Wyatt, rev., Harvard University Press.

———. 1925. *Homer: Iliad*, vol. II: books 13–24 (*Loeb Classical Library* 171), W.F. Wyatt, rev., Harvard University Press.

Robinson, J., J. Harding, and J. Vouillamoz. 2012. *Wine Grapes: A Complete Guide to 1,368 Vine Varieties, Including Their Origins and Flavours*, Ecco.

Stavrakaki, M., and M.N. Stavrakakis. 2017. *The Cretan Grapes*, Tropi Publications.